U0299333

臻藏美墅

BLOOMING VILLAS

TO SEE A WORLD IN A HEAVENLY MANSION

一府一天下　一墅一世界

DAM 工作室 主编

華中科技大學出版社
http://www.hustp.com
中国 · 武汉

PREFACE

As the symbol of success, wealth and status, the top villa represents people's desire for material life. It is the foundation of the culture context and the starting point of the legacy with non-renewable resources, mature commercial services, the exclusive supporting system, humanized requirement and technology concern. Under the kind invitation of Huazhong University of Science & Technology Press, I am honored to preface for the book *Blooming Villas*, hoping to inspire the young designers and show directions for the students, sharing the experience with readers.

The origins of geomantic omen in China can be traced back to the original human hunting period, then the human already knew to choose apricus caves as the dwelling place. In the period of ancient Fuxi, it appeared "figure in Yellow River, book in Luo River" legend, Fuxi accordingly evolved it into the Eight Diagrams, gradually turned it into *The Book of Changes*. In Qin dynasty there appeared the ideas of Geographical Position and King's spirit so that the First Emperor of Qin built magnificent tomb for himself, including the unearthed terracotta warriors. In Wei and Jin period there appeared the masters of geomantic omen such as Guan Luo and Guo Pu with books like *Guan Shi Geography Guidance* and *Burial Books*. In Tang dynasty, geomantic omen was divided into two factions, that is, the situation and regulating vital energy which also merge together. During the Ming and Qing period, geomantic art obtained great development, many masters and books emerged in endlessly...

Nowadays, geomantic omen has been spreaded to the world. There are 17 universities in America opening the relative courses, including the most famous Berkeley University in California. In Japan, there are 110 universities offering geomantic omen courses. The upsurge runs through many countries such as Spain, Australia, Singapore, Thailand, Malaysia. The geomantic omen in Hong Kong, Macao and Taiwan is always very popular. I started my world tour academic speech activities at the return of Hong Kong in 1997. I have given over 100 lectures at home and abroad. The active audience impressed me, among them there are designers, architects, developers, lawyers, brand suppliers and so on.

With the development of reform and open up, under the background of the social and political environment and academic atmosphere, plenty of geomancy learners appear for people require a more qualified living environment. However, under the right conditions, it is not responsible for people to completely deny or accept the geomantic omen. In recent years, I have heard from many readers who give a high evaluation to my academic views and expecting better works from me. Now I share my experience in geomancy villa design with readers.

Different with the common residence, the villa is easier to be influenced by the surroundings. Therefore, it is necessary to judge the geomancy of the villa in the respect of inside and outside environment. By and large, the outside environment influences the geomancy of the villa more greatly than the inside environment. Therefore, lighting, greening and wellspring are the basic elements in the surroundings when choosing villas. On this basis, the interior structure needs to be analyzed to select the most suitable villa.

The geomancy theory concludes the natural laws such as the relative astronomy, meteorology, geography knowledge and life experience. It is said that the geographical environment such as fronting water with hills on the back and southern exposure are good for the resident's health and wealth. The shape also plays a role in the fortune of the owner, among which the square villa is the best, magnificent and restrained. The rectangular villa is regular and suitable for everyone. Without sharp edges and corners, the circular villa represents reunion which is also operable.

The living room is the place for meeting and entertainment, the main activity space for the family. Therefore, it would better be set in the center of the villa and should be the largest area in the space. Besides, it needs more light and ventilation to benefit the family.

Because of the large area, most of bedrooms in the villa are designed larger than that of the common residende. Living in the large bedroom will consume human's energy, immune system decline, even illness. Therefore, the bedroom should not be much too large, from 20 to 30 square meters will be better. The quadrate shape is the basic element for the bedroom which can't be polygon. If the loft is used as the bedroom, the hypotenuse in the roof is easy to mislead people visually to burden people, resulting in illness or accident. Therefore, The loft is suggested to be used as the study or the storage room.

Genarally, the villa has two floors or above, therefore, we should take the whole layout into consideration, especially the design of the bathroom. In the geomantic omen, the upper level and the below level have a close relationship. Every level will set a single bathroom. The upper bathroom cannot above the below bedroom, for it will introduce the bad things to the below bedroom to harm the health and luck of the resident. Further more, the stairs are regarded as the key place of vent, not only acting as the passage but also influencing the resident's fortune. The stairs should lean against the wall to avoid dividing the open space, resulting in visual and psychology trouble.

The garage should be designed to meet the rules of convience, watertightness, ventilation and avoid environment pollution. Besides, the garage should not be designed under the bedroom, especially the same side with the window or balcony of the bedroom. The waste gas will enter the room to harm people's health and wealth. To diminish the harm of the waste gas, more trees such as pines should be planted in the entrance of the garage.

Many villas will build the rockwork and fountain to beautify the environment while this is not a good choice for the entrance of the villa. The best places are both sides on the door far away from the house so that people cannot hear the sound of the fountain.

The water is characterized by Yin with a heavy humidity. It is necessary to pay attention to the swimming pool's setting. The swimming pool can't be set too close to the house, which will increase the humidity which is bad for the resident. Besides, the swimming pool can be curved to embrace the house visually, letting the owner see the swimming pool once opening the window, taking advantage of the swimming pool's energy.

After years' pratice reserch, I find that it is using the nature environment and superiority to initiatively plan the house that benefits not only the resident's health but also the family's fortune. Here I want to appreciate my clients who allow me to study and practice in their residences. In a word, as a multidisciplinary subject, villa geomancy design has a close relationship with other subjects. It is a strick task to perfect this subject over time to meet human's requirement. Finally, looking forward to readers' correction for my limitation.

Jacky Chan
Summer, 2013
Hong Kong

序言

顶级别墅作为成功、财富、地位的象征，代表了人们对物质生活的最高追求。它拥有不可再生的稀缺资源、成熟的商业服务、专属的配套体系、人性化的需求及科技关怀，是文脉的根基，传世的起点。受华中科技大学出版社的盛情邀请，笔者非常荣幸能为《臻藏美墅》作序，希望通过笔者在风水设计方面的实践，既能为年轻设计师传授经验，又能给设计类学生指引方向。在此抛砖引玉，与有缘人共勉。

中国风水的起源，可以追溯到原始人类的狩猎时期，那时的人类就已经懂得选择避风、向阳的洞穴作为居所；上古伏羲氏时，出现了“河出图，洛出书”的传说，伏羲依此而演变成八卦，后逐渐演变成《易经》；秦朝时有了地脉和王气观念，秦始皇给自己修建了宏伟的陵墓，现代出土的兵马俑只是其中的一部分；魏晋时期出现了像管骆、郭璞这样的风水宗师，有《管氏地理指蒙》《葬书》等典籍问世；唐代，风水分为形势和理气两大派别，两派之间既相互碰撞又相互融合；明清时期风水学得到了极大的发展，各派别人物和典籍层出不穷……

而今，中国的风水学已被传播到世界各国，美国目前已有 17 所大学开设风水学课程，其中包括加利福尼亚州最著名的伯克利大学。日本已有 110 所大学开设风水学课程，韩国更是不计其数，西班牙、澳大利亚、新加坡、泰国、马来西亚等国家均掀起一股风水学热潮。香港、澳门、台湾地区的风水学一直倍受青睐，笔者正是 1997 年香港回归之时开始世界巡回学术演讲活动的，目前在国内外的演讲已超过 100 场，让笔者感触最深的是港澳台地区听众的活跃，他们当中有设计师、建筑师、地产商、律师、品牌供应商等。

随着改革开放的不断深入，国内的社会环境、学术气氛得到进一步改善，人们对居住环境的要求不断提高，因此涌现出一大批风水学者。但是，在当前形势下，对风水学抱着简单的全盘继承或全盘否定的态度都是一种不科学、不负责的态度。近年来，笔者收到了大量国内外读者的来电、来函，对笔者主张的学术观点给予很高的评价，并期待笔者有更多更好的作品呈现，现将别墅设计方面的经验整理出来，以飨读者。

别墅与普通住宅不同，它更容易受到周围环境的影响。因此，在衡量别墅的风水时，必须从内环境和外环境两个方面进行判断。总的来说，外环境对别墅风水的影响大于内环境。因此，在选择别墅时，应重点考察周围的环境，采光、绿化和水源是最基本的元素。在此基础上再进一步分析别墅的内部结构，选择最利于宅主运势的别墅。

风水学理论总结了与建筑相关的天文、气象、地理等自然规律和相应的知识与生活经验，其认为背山面水、坐北朝南的地理环境符合别墅最佳选址的条件，也最利于财气的聚集与居住者的身心健康。别墅的形状对宅主的运势也会产生影响，正方形的别墅稳重大气，是最佳的选择。长方形的别墅中规中矩，适合所有人。圆形的别墅，由于没有尖锐的棱角，象征着团圆与吉祥，也是可以使用的。

客厅是家人聚会和接待宾客的地方，是家庭的主要活动空间，所以其位置最好处于别墅的中心，而且从面积上来讲，应是整栋别墅中空间最大的。并且客厅需要有充足的采光与良好的通风，只有这样才能用充足的阳气带动家运的旺盛。

由于别墅面积较大，大多数别墅的卧室都设计得比普通住宅的卧室大得多。其实，长期居住在面积较大的卧室中，人体的能量相对来讲会消耗得更多，也更容易出现精神不佳、免疫力下降等状况，严重者还会出现疾病。因此，别墅的卧室不可盲目求大，最好控制在 20~30 平方米。形状方正是卧室最基本的要求，不能是多边形或是有斜边出现。如果将阁楼作为卧室，屋顶的斜边很容易造成视觉上的错觉，而由此斜边构成的多边形卧室，也会使人的精神负担增加，容易发生疾病或意外。所以，如果想有效利用别墅的阁楼，建议用来做书房或储物间。

别墅一般有两层或两层以上，因此，在进行房屋功能规划以及装修时，不能单一地考虑每一层的设计，而应该综合考虑整栋别墅的格局，尤其是卫生间的设计。在风水学中，上下两层楼之间有着密切的联系。一般别墅的每层都会设有单独的卫生间。此时就必须要注意，上一层的卫生间最好不位于下一层的卧室之上，这样会导致楼上的污秽之气流散到楼下的卧室中，从而影响到居住者的健康和运势。再者，楼梯被视为别墅中解气和送气的关键之处。因此，别墅楼梯不仅是上下楼之间的通道，更关系着宅主的运势。在设置楼梯时，要尽量选择靠墙的位置，如果将楼梯设置于一楼房屋的正中央，这样不仅打破了空间格局，使空间一分为二，还使房屋的空阔感顿失，造成视觉和心理障碍。

在设计别墅车库时，首先要满足宅主进出的方便，其次要注意防水和通风，还需做好清洁工作，防止尘土对环境的污染。另外，车库最好不要设在卧室下方，尤其是卧室窗口或阳台与车库位于同一面，这样废气很容易从阳台或窗户进入室内，影响人的健康，对财运也会有影响。在无法改变位置的情况下，应在车库出口处多种一些绿色的植物，以减少汽车尾气的影响。

为了达到美化环境的目的，许多别墅都会建造假山和喷泉，如果选址不对则会带来很大的影响。一般来说，在别墅的大门口建造带有喷泉的假山并不是大吉的。如果一定要在此建造，则需建在离大门有一定距离的地方，以在室内听不到喷泉的声音为宜。

水，属阴性，而且湿度相对较大。故游泳池的设置与使用应该极为注意。游泳池不宜距房屋过近，其易增加空气湿度，影响居住者的身体健康，进而影响气运。另外游泳池也可设计为曲线形，“8”和葫芦形泳池最为常见，这样在视觉上成水环抱房屋状，也符合面水的选址条件。让宅主开窗就能看到泳池，以有效利用泳池的风水能量。

笔者通过多年的实践发现：合理地利用自然环境与自然优势，能动性地规划居室布局，不仅有利于居住者的身心健康，而且对家庭运势也有很重要的影响。在此，我想感谢所有给予我信任的客户，感谢他们容许我在其居所中进行学习和实践。总之，别墅风水设计作为一门综合性学科，与其他学科均有紧密的联系，怎样与时俱进，不断完善这门学科，为现代社会的人类需求作出更大的贡献，需要我们共同努力。因笔者水平有限，不妥之处在所难免，望有缘于此书的读者批评指正。

陈杰
字浩龙
癸巳年夏于香港

FOREWORD

前言

History of the Former Han Dynasty · Dong Zhongshu Biography recorded, better action than irresoluteness. Like all mortal beings in the big world, walking over the busy city and suffering from the annoyance, people live by the water in the city to embrace the nature rather than yearn for the countryside life. Like the flower in mist and the cloud-castle, everything is so wonderful and surprised over time.

The saying goes, "The house is not the simple construction but people's inner building." So are villas. All the residences start from the heart of the residents to create a humanized space, borrowing the beauty of the nature to satisfy the residents' persuing for the good things, making a dialogue between the rivers and the soul to intoxicate people in the legend.

The conatation of villas lies on the silence, the nature and the luxury space. However, villas are not only owned by the rich, but also possesses the elegant and dignified atmosphere, integrating the strength, wisdom and temperament to read the life in a cultural perspective. The space can be luxury and magnificent without ostentations, showing the cultural spirit in the layout.

In the villa, one can not only live in the bustling but also attain peace in the heart. The starlight night is not the ending but the beginning. The fantastic lamps, the glorious space, the twinkling night sky, all build a world full of peace and warm.

We still yearn for the class act while it is a hard act to follow. Open the book *Blooming Villas*, to persue the beauty of the housing.

《汉书·董仲舒传》里曾有言曰：临渊羡鱼，不如退而结网。于凡嚣之中，如万万人一般，穿梭于热闹的都市，承受千万人之烦闷，歆羡湖畔山间的鱼鸟之乐，憧憬南山的东篱之情，不如于都市择山水而居，于郊外享自然之灵韵。既有雾里看花之朦胧，亦有空中楼阁之梦幻，一切都如此美好，是惊喜，也是心中的“小确幸”！

有人曾说“房屋不是在地面上的简单搭建，而是从人的内心筑起”。别墅，也不例外。不论是古代帝王的行宫、将相的宅邸、商贾巨富的庄园，还是如今都市中的艺墅园，抑或是青山绿水中的度假山庄，无一不是从居住者的内心出发，打造人性化的尊享空间，臻萃山之诗情、水之画意，满足居住者对美与爱的追求，实现碧水与心灵的吟唱与对答，让居者醉心于这遗世的传奇。

别墅之意不仅在其静雅之境、山水之情，更在于其豪与奢的气度。但这也绝不是说别墅专为多金人士所有，真正的别墅要有高贵典雅的气质，要融力量、智慧与性情于一体，更要有以人文精神品鉴生活的眼界。雍容华贵会有，气势恢宏也有，但这绝不是虚张声势，更不是设计的浮夸，而是意在于铺排之中彰显人文气度，在气度中描绘居者之风度。山水在院，气质盈室，情浮于空，爱存于心，何人不喜?

据此美墅，进可享世间繁华，退可得内心逸静。星光下的夜晚，不是结束，而是开始。巧设的灯饰、满室的辉煌、闪亮的夜空，洞明无数世事，曾经历尽的沧桑，此刻，皆已幻化，唯留满心的安宁与温暖。

“高山仰止，景行行止”，虽不能尽至，然心向往之。对于居室之美的追求亦是如此，别墅的闲情，爱与心的邂逅，从《臻藏美墅》开始！

CONTENTS | 目录

Poly 12 Oak Manor House 008
保利十二橡树庄园别墅
Shuiyue Dragon Bay Villa No. 61, Foshan 018
佛山水悦龙湾61号别墅
Jindi Green Shore Ze Park Villa Sample B 030
金地格林春岸·泽园别墅B户型
Taihe Hongyu G04 Villa Show Flat, Beijing 044
北京·泰禾红御G04别墅样板间
Shenzhen Beachfront Duplex Apartment 056
深圳滨海复式公寓
Discovering California–Shenzhen Tianyu Mountain Duplex 064
情迷加州–深圳天御山复式
Changzhou Daming City 072
常州大名城
Mountain Top Residence 078
山顶大宅
The Lemon House 084
香港大宅
Abbots Way 090
Abbots Way别墅
The Lighthouse 65 098
灯塔65
Joc House 104
Joc别墅
House Refit 112
Refit别墅
Apple Bay House 118
苹果湾别墅
Seaview House 124
海景别墅
Cultural Holiday Villa 128
人文度假别墅
Taihu Tianque, Suzhou 136
苏州太湖天阙
Huizhou Golf Villa 152
惠州高尔夫别墅
Winds to Aegean Sea 158
风向爱情海
New Canaan Residence 164
新迦南别墅

West Lake Hills Residence 172
西湖山别墅
Norwegian Official Residence 180
挪威官邸
Kowloon Bay Villa 184
龙湾别墅
Casa LC 192
LC 别墅
House in Rocafort 198
罗克福德别墅
Xiamen Bali Villa 204
厦门巴厘香墅
American Style FULL-SUN Villa 210
府尚别墅·美式
Neoclassical Style FULL-SUN Villa 218
府尚别墅·新古典
Spanish Style FULL-SUN Villa 224
府尚别墅·西班牙风格
Modern Style FULL-SUN Villa 230
府尚别墅·现代风格
Yihu Beautiful Home 236
怡湖·美家
Santa Monica Canyon Residence 242
圣塔莫尼卡峡谷屋
Huizhou Zhongzhou Central Park 248
惠州中洲中央公园
West Hill Whispering Woods 264
西山林语
Chinese Tuscan Dream 278
中国人的托斯卡纳梦
A16 Villa Show Flat, Nansha Xinghe 284
星河南沙A16别墅样板房
New World Dream Lakeside Villa 294
新世界·梦湖香郡
Vuitton Town of Hot Springs Villa-B 302
威登小镇温泉Villa-B
Vuitton Town of Hot Springs Villa-C 308
威登小镇温泉Villa-C
Cang Hai Villa No. 1 Townhouse Unit A1 326
苍海一墅联排别墅A1户型
The Feeling of "Play in the Art" 334
游于艺的Feeling

Poly 12 Oak Manor House

保利十二橡树庄园别墅

◎ Design Company: IEA Design
◎ Designer: He Xuan
◎ Location: Wuhan, Hubei
◎ Area: 400 m^2

◎ 设计公司：武汉艾亿威装饰设计顾问有限公司
◎ 设计师：何璇
◎ 地点：湖北武汉
◎ 面积：400 m^2

The owner expects a young and easy space without ostentation. The western European style architecture is influenced by the Moorish culture. The project conbines the Middle Eastern style and the European style, for example, the doorway is the Middle Eastern style while the main background of the living room and the bedroom is the western European style echoing with the facade.

The space is accented with simple white lines, extensive bright wallpaper, black marquina and block floor. All the lamps can be controlled intelligently so that it is easy for the owner to operate. The collocation of colors is very important in the space. The space includes the bright colors such as the bright yellow and the light green matched with some dark colors such as brown and blue, accented by the golden color, attaining a perfect balance.

本案业主希望营造一种较为活泼、轻松的氛围，设计无需刻意追求豪华，不必过分张扬。历史上摩尔人的文化对西欧的建筑风格产生了很大的影响。本案设计将中东的风格和欧式的风格相结合，如门洞的设计就采用了中东的设计形式，而客厅和卧室的主背景则又运用了欧式的设计手法，也呼应了欧式的外立面。

空间采用了简洁的白色线条和大面积鲜艳、明快的墙纸以及黑白根石材、拼花地板等材料。室内所有的灯具都可智能调控，这样更易于操作。空间色彩的搭配尤为重要，故空间选用了明黄色、淡绿色这样明快而又清新的色彩，搭配深一点的咖啡色和蓝色来压住跳跃色彩的浮躁，并采用金色作为点缀。

Shuiyue Dragon Bay Villa No. 61, Foshan

佛山水悦龙湾 61 号别墅

◎ Design Company: Shenzhen Hoverhouse Project Design Co., Ltd.
◎ Designer: Kevin Chen
◎ Location: Foshan, Guangdong
◎ Area: 500 m^2
◎ Main Materials: Stone, Ceramic Tile, Stainless Steel, Leather, Cloth Art

◎ 设计公司：深圳雅典居设计
◎ 设计师：陈昆明
◎ 地点：广东佛山
◎ 面积：500 m^2
◎ 主要材料：石材、瓷砖、不锈钢、皮革、布艺

HEPBURN

Star ★ Collection
PRESLEY
Star ★ Collection
HEPBURN
PRESLEY
HEPBURN

The project is a modern style single family villa, creating a dignified, avantgarde and Oriental living space through the method of dark and light colors contrast.

The living room is light color, achieving the visual transition by the different colors. The kitchen is half open, revealing the warmth and harmony of the home, occasionally adding a simple and fashional temperament to the space. The main bedroom uses two main colors which are matched well to show an open and bright space. The underground floor is skillfully designed family area, divided into billiards room, bar, chess and card room, video room, creating an elegant and peaceful atmosphere.

Basement Plan / 地下层平面图

本案是具有现代时尚气息的独栋别墅，设计运用动中取静的手法，并通过深浅对比，勾勒出一幅尊贵、时尚兼具东方情怀的家居生活蓝图。

客厅以浅色调为主，深浅不一的色调实现了视觉上的过渡。厨房半开放式的格局，既透露出家的和谐与温馨，又在不经意间突显出空间简约、时尚的气质。主卧室采用了两种主要的颜色，搭配简单、大方，更衬托出空间的开阔与明亮。地下一层是家庭的活动空间，空间依据活动类别划分为台球室、酒吧、棋牌室、影音室，设计精致且不着痕迹，营造出典雅、祥和的气氛。

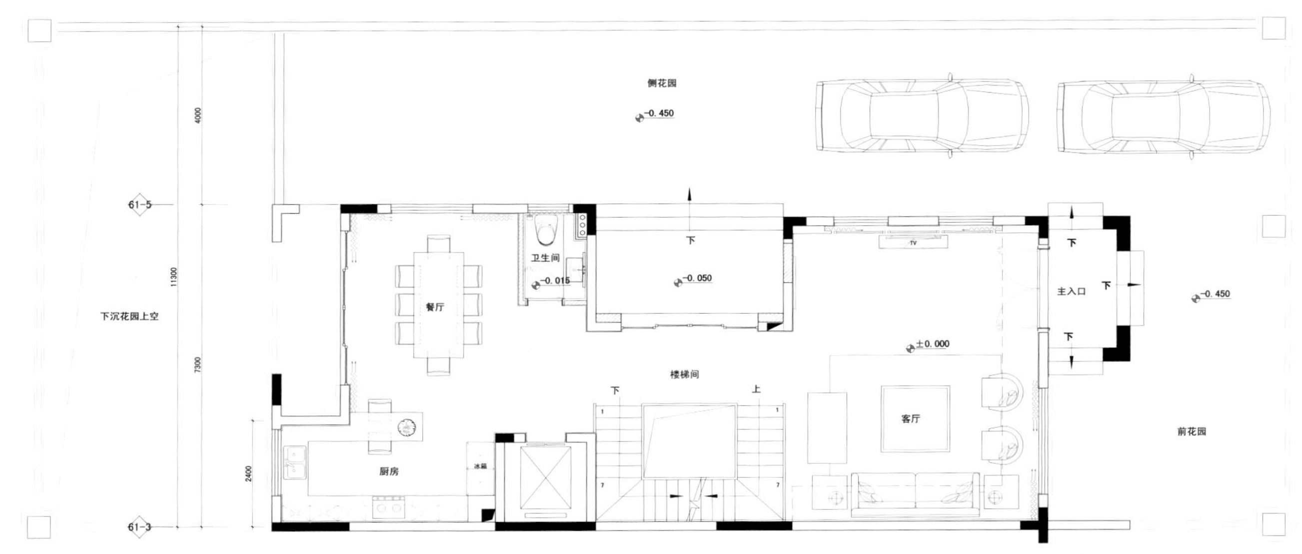

First Floor Plan / 一层平面图

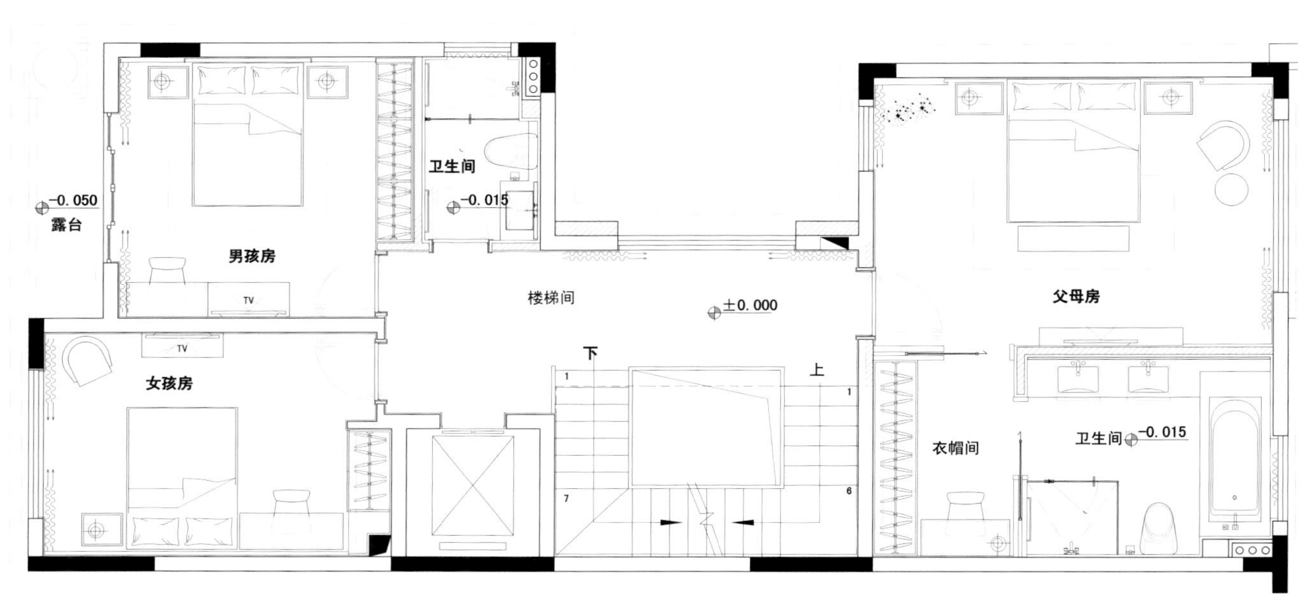

Second Floor Plan / 二层平面图

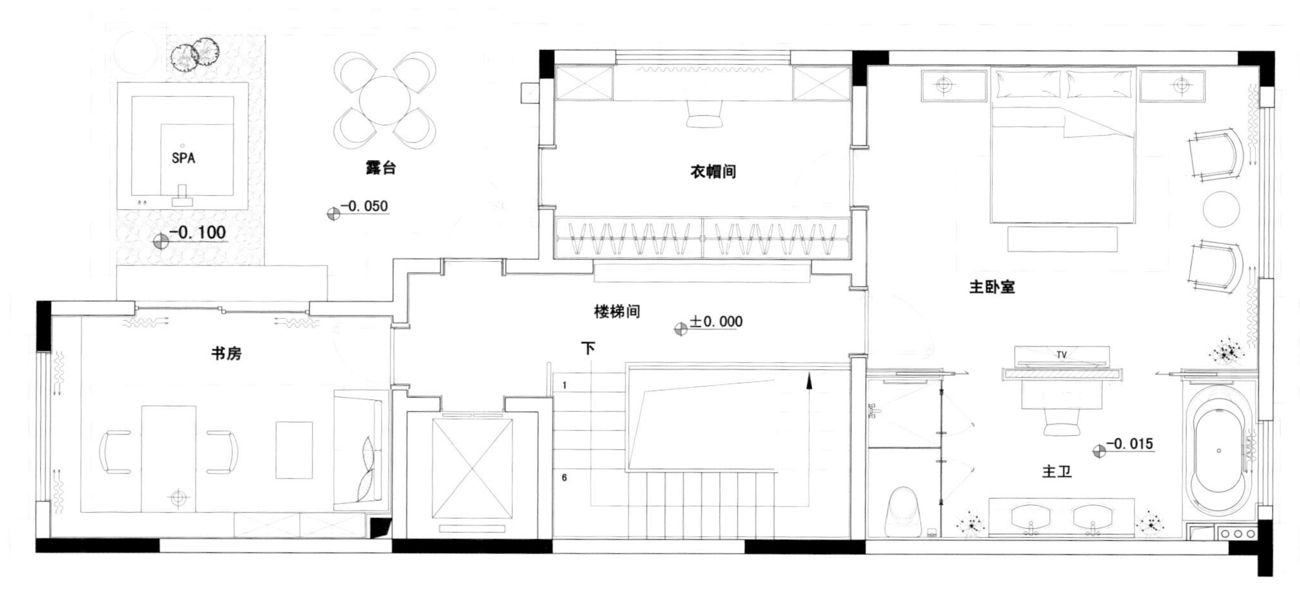

Third Floor Plan / 三层平面图

Jindi Green Shore Ze Park Villa Sample B

金地格林春岸 · 泽园别墅 B 户型

◎ Design Company: Much Art & Design
◎ Designer: Chen Yi, Zhang Muchen
◎ Location: Wuhan, Hubei
◎ Area: 500 m^2
◎ Main Materials: Croatia Stone, Fancy Brown Stone, Solid Wood Floor, Emulsioni Paint, Wallpaper, Wood Beam, Metope, Wood Veneer, Mosaic, Stained Glass

◎ 设计公司：北京睦晨风合艺术设计中心
◎ 设计师：陈贻、张睦晨
◎ 地点：湖北武汉
◎ 面积：500 m^2
◎ 主要材料：克罗地亚石材、热带雨林啡石材、实木地板、乳胶漆、壁纸、木梁、墙面、木饰面、马赛克、彩色玻璃

Upon entering, the small stream, the green pots, the leisure chair, the lotus, all become part of the Oriental courtyard.

In the living room, people can gleamingly see the study in the second floor under the yellow chanderlier, full of books and luxury. The study is also like a viewing deck where people can see the living room to appreciate the layout and items.

Walking over the lobby, there is a dining room which is a multi-functional space integrating dining room, kitchen and bar area. The designer breaks through the three spaces of the original architecture to divide the area reasonably. The center is a dining area with bar area and kitchen beside. There is an buffet between the sliding

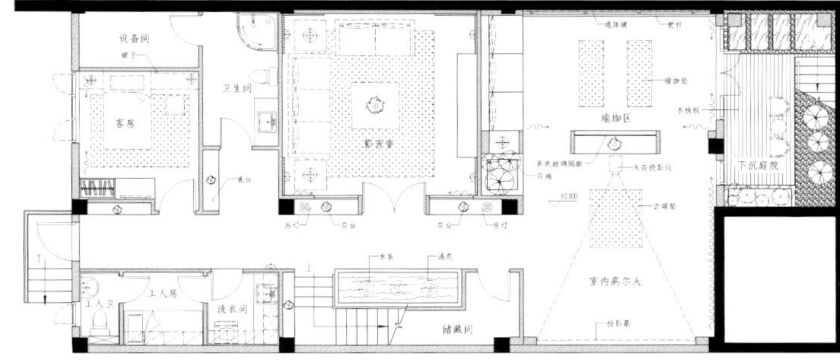

Basement Plan / 地下层平面图

doors in the kitchen. The bar echoes with the kitchen, and the kinetonema is designed on both sides of the bars, which not only conforms to the eastern symmetry rule, but also is western life style.

To show the Thai feature, the space includes the table and chair combining the Thai design element with European style. The whole collocation, no matter the hard surfaces or the soft installations, every decoration, talk with each other.

一入庭院的小门，涓涓的流水、绿色的陶罐、舒适的休闲椅、含苞的荷花，具有东方情调的庭院瞬间展现在眼前。

在客厅，可以仰望二楼的书房，透过精妙的格栅可以隐约看见书房中的情景，书柜、陈设在泛黄水晶灯的映照下，散发的不仅仅是书香味，更是奢华的皇家贵族气息。如观景台一般的书房亦可回望客厅，欣赏室内精致的布局与物件。

穿过走廊就是餐厅，它是一个集餐厅、厨房、酒吧区于一体的多功能空间。设计师在结构上稍动手笔，将原建筑的三个空间打通，又进行了合理的区域划分。就餐区位于中间，两侧分别是酒吧区和厨房。厨房设置有推拉门，两个门中间是一个岛台。吧台与厨房相对，动线设置于吧台两边，既符合了东方的对称法则，又具有了西方的生活情调。

为了突显泰国的特色，设计师专为餐厅设计了特别的餐桌、餐椅，既将泰国的设计元素融入家具中，又不失欧洲味道。整个空间的元素搭配，无论是硬装还是软装，似乎每个布置、每个装饰语言，都在对话。

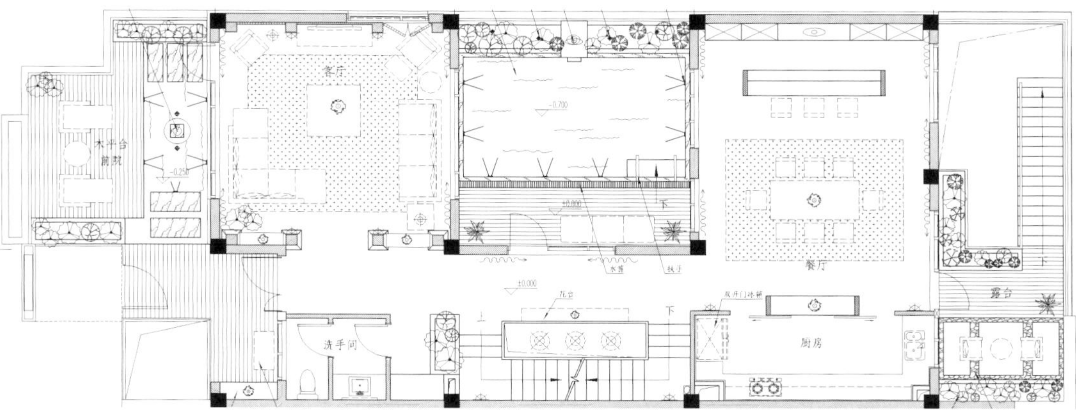

First Floor Plan / 一层平面图

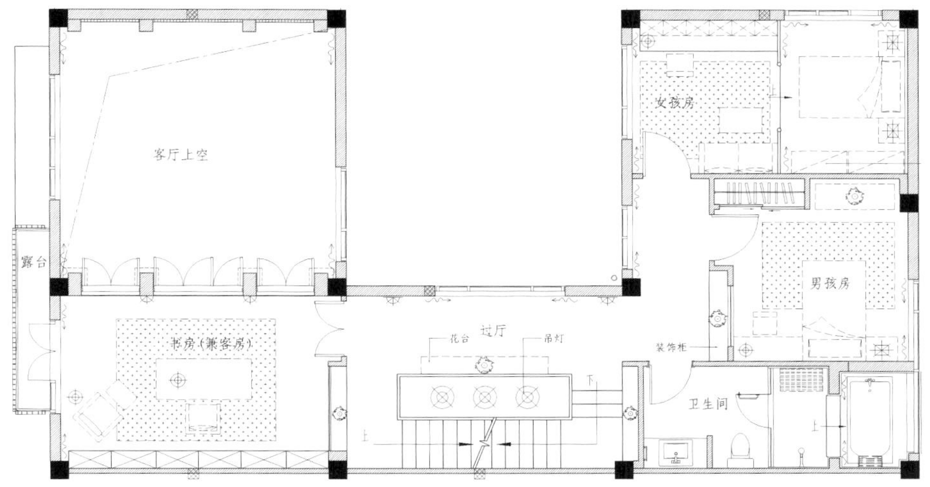

Second Floor Plan / 二层平面图

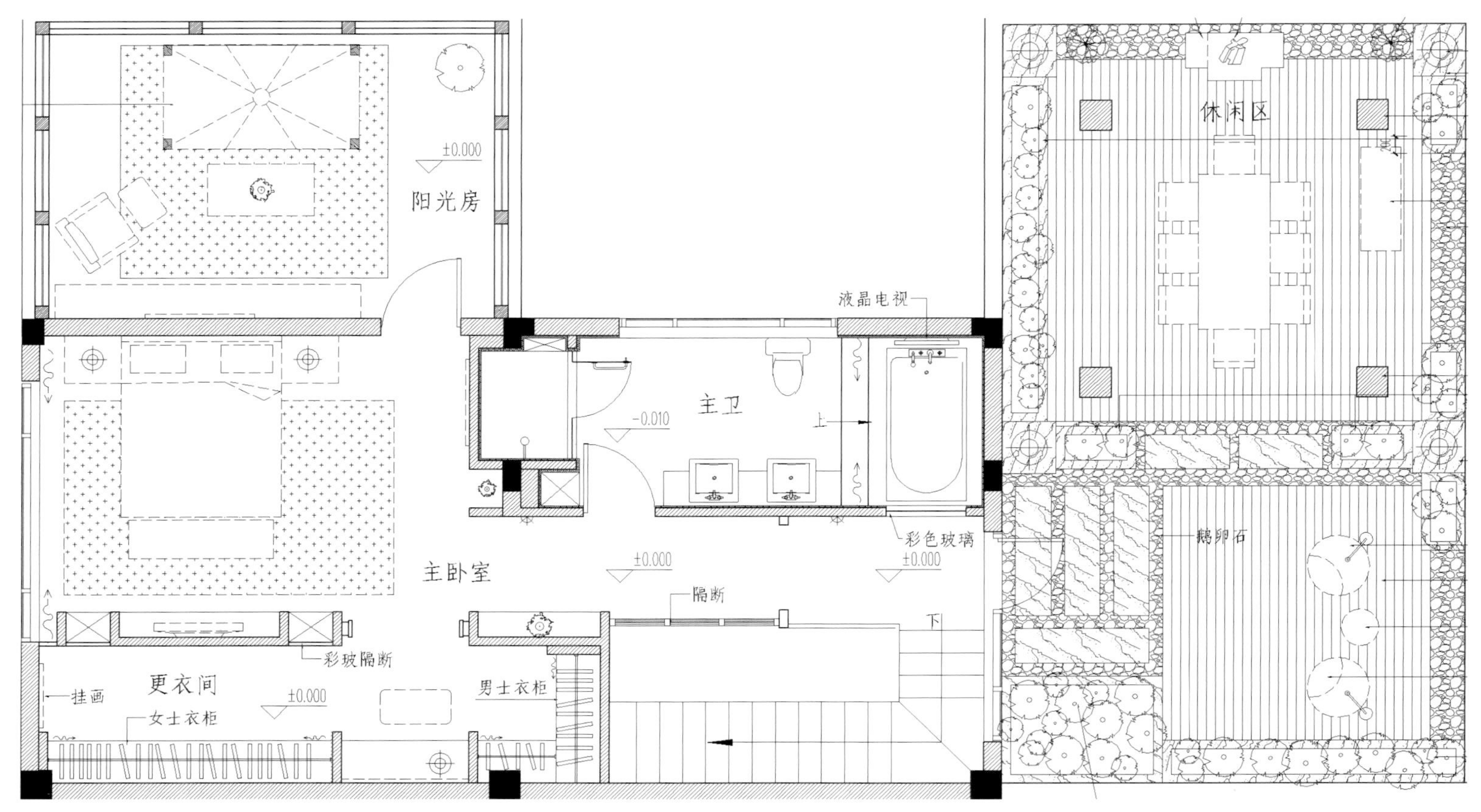

Third Floor Plan / 三层平面图

Taihe Hongyu G04 Villa Show Flat, Beijing

北京·泰禾红御 G04 别墅样板间

◎ Design Company: Gotomaikan International Limited
◎ Designer: Xu Shaoxian
◎ Location: Beijing, China
◎ Area: 1,479 m^2

◎ 设计公司：香港五斗米馆国际有限公司
◎ 设计师：徐少娴
◎ 地点：中国北京
◎ 面积：1 479 m^2

The project uses the French Neoclassicism design method to merge the French elegance element into the neoclassical kingly style, bursting out a special romantic feeling. The progressive colors and shapes, modest and tasty, add more romantic elements to the space with the gold embellishments.

The first floor in the living room uses stone pattern to modify the artificial trace with the stone's natural texture and color, unreservedly showing the luxury and taste in the living room and dining room. The whole interior space is fluent and bright but changeable, maintaining the continuity of the space.

The basement mainly focuses on the entertainment. The mosaic floor, the long bar, the antique mirror decorated metope with various point lamps, make the light and the shadow interlace and merge toghether to become colorful and elegant in the understated dignity. Also, the colorful carpet lights the whole space.

The seond floor includes four bedrooms. The study is designed in the main bedroom, through which we can see the whole scene in the living room. The bedroom is warm color, showing dignity by the French wood bed and the gold chanderlier. The dark flower pattern curtain and the beige sofa represent the understated life attitude of the owner.

本案采用了法式新古典的设计手法，将法式特有的典雅融入新古典的高贵，从而迸发出另一种独特的浪漫风情。层层递进的色彩与造型，稳重而极具品位，金色的点缀，让空间无意间增添了更多浪漫的元素。

在设计上，客厅地面采用石材拼花，用石材天然的纹理和自然的色彩来掩饰人工的痕迹，使客厅和餐厅的奢华与品位毫无保留地呈现。别墅整体的室内空间一气呵成，通透明亮，但又极富变化，保持了空间的连续性。

地下室则以娱乐空间为主，马赛克地面、长条的吧台、仿古镜装饰的墙面配上各种点光源，使空间中的光与影互相交错又互相融合，绚丽而不失庄重，低调中散发优雅。色彩绚丽的地毯，使整体空间沉稳而不失生动。

二楼设有 4 个卧室，书房在主卧中，可以通过书房的挑台看到整个客厅的景象。卧室主要以暖色调为主，纯正的法式大木床，金色的水晶吊灯，处处彰显着高贵。深色印花的窗帘和米色的沙发躺椅则传达出主人低调、内敛的生活态度。

Shenzhen Beachfront Duplex Apartment

深圳滨海复式公寓

- Design Company: YuQiang & Partners Interior Design
- Photographer: Wu Yongchang
- Location: Shenzhen, Guangdong
- Area: 483 m^2
- Main Materials: Wood Veneer, White Matte Paint, Pure White Artificial Stone, Black Sand Steel

- 设计公司：于强室内设计师事务所
- 摄影师：吴永长
- 地点：广东深圳
- 面积：483 m^2
- 主要材料：木饰面、白色哑光漆、纯白人造石、黑砂钢

First Floor Plan / 一层平面图

Second Floor Plan / 二层平面图

Campbell's
CONDENSED
TOMATO
SOUP

The project as a duplex seaside mansion with fantastic seaview, it's has an incomparable landscape advantage. It becomes a very important element that bring the most advantaged landscape into the interior to make the layout more open and transparent in the floor plan.

The first floor is a public area includes the living room and the dinning room which open to each other with communications, invisibly expand the space with more landscape. The main bedroom and the private leisure area are designed in the second floor, highlighting the easy and tranquil atmosphere compared with the dynamic in the first floor.

The project takes advantage of many wood elements, silk, linen and stones with natural texture, representing an easy, leisure and cordial seaside mansion temperament.

本案为拥有一线海景的复式滨海大宅，有着无可比拟的景观优势。在平面布局过程中，如何使格局开敞、通透，将其最具优势的景观引入室内，成为设计的关键。

一层为公共区域，客厅与餐厅的开敞格局，使两个相对独立的空间形成互动，为空间引入更多景观的同时，也于无形中扩大了空间感。主卧室及相对私密的休闲区设在二层，相对于一层的“动”，二层更多的突显了轻松与安静的氛围。

在材质的选用与搭配上，本案将大量的木元素与丝、麻及部分有着天然肌理的石材相搭配，营造出轻松、休闲又不失亲切的滨海大宅气质。

Discovering California-Shenzhen Tianyu Mountain Duplex

情迷加州 – 深圳天御山复式

◎ Design Company: Shenzhen Hongyiyuan Architecture Interior Design Co., Ltd.
◎ Designer: Zheng Hong, Xu Jing
◎ Photographer: Justin
◎ Location: Shenzhen, Guangdong
◎ Area: 280 m^2

◎ 设计公司：深圳鸿艺源建筑室内设计有限公司
◎ 设计师：郑鸿、徐静
◎ 摄影师：Justin
◎ 地点：广东深圳
◎ 面积：280 m^2

文伟阁

The charm of the simplified European style is that its shape, texture, incising and color are concise, fine and smooth, emitting an everlasting fragrance. Influenced by the American style, the temptatious wood colors the gentle space, surprising us with an enthusiastic romance.

To satisfy the different home decoration intention of the owner, the designer magically conbines the simplified European stylean hard surfaces with American style soft installations, representing the elegance of the hostess and the enthusiasm of the host. The space is luxury and magnificent, while free and casual. The classic elements and the nostalgia breath along with fresh sea wind show the art infection and the love for California.

简欧风格的迷人之处在于其造型、纹理、雕饰和色调的简约与细腻，耐人寻味处散发出亘古久远的芬芳。而在美式风情的渲染之下，淡妆浓抹的木色，也为细腻温婉的硬装注入七彩阳光，砰然撞击着视网膜的最大阈值，使静美的空间多了一份热情与浪漫。

为了满足业主迥然不同的家装意向，设计师巧妙地将硬装简欧风与软装美式风结合，简约娴雅的硬装设计突显女主人的高雅，鲜亮华丽的软装则象征了男主人的热情爽朗。空间华贵大气，又不失自在、随意。古典元素的运用，增强了空间的其艺术感染力，悠远的怀旧气息，夹带着西海岸清新的海风，氤氲着太平洋彼岸的加州情怀。

欢迎！
经过简单的设置您就可以启动LG智能电视。

Changzhou Daming City

常州大名城

◎ Design Company: Pure Charm Space Design Organization
◎ Designer: Lin Xinwen
◎ Photographer: Shi Kai
◎ Location: Changzhou, Jiangsu
◎ Area: 530 m^2

◎ 设计公司：福建品川装饰设计工程有限公司
◎ 设计师：林新闻
◎ 摄影师：施凯
◎ 地点：江苏常州
◎ 面积：530 m^2

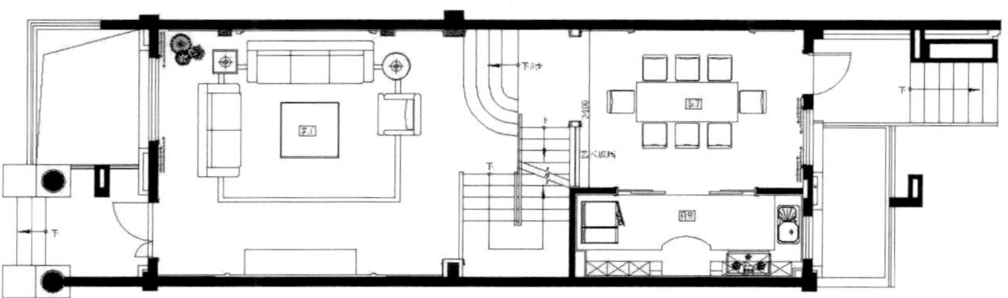

First Floor Plan / 一层平面图

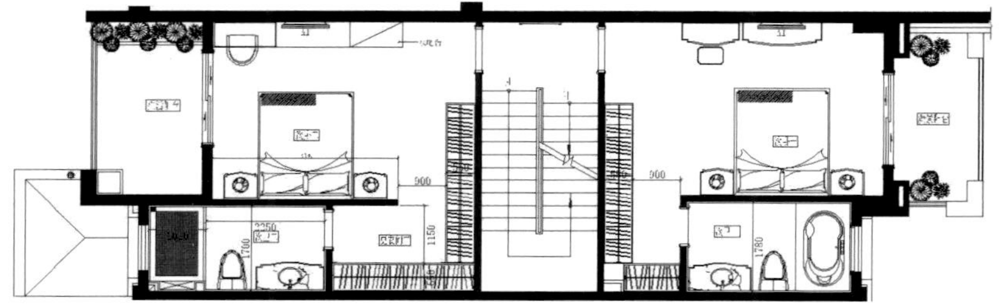

Second Floor Plan / 二层平面图

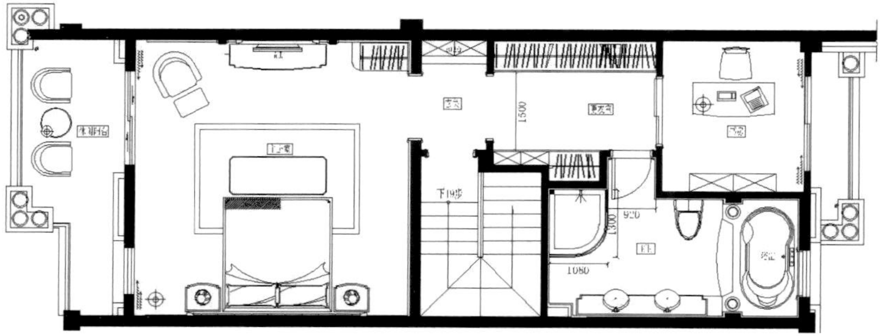

Third Floor Plan / 三层平面图

The classic European on style adds a breath of elegance and dignity to the space. The huge furniture in the living room is not cumbersome but enhances the comfort as well as satisfies the visual requirement. The ivory color under the gentle light creates the warmth of the home. The dining room is the "stomach" of the home, with a half open door design in the whole European style. The furnitures in the bedroom take advantage of the classical elements, like the red wine in Bordeaux left bank, bringing people a sincere cultural feeling warmly. The study is simplified European style, showing a cultural ambience and the owner's cultural deposits.

经典的欧式设计风格，带来的是一种典雅、高贵的空间气息。客厅在软装上采用大体量的家具，不仅不显笨重，反而在满足视觉需求的同时，也增强了舒适感。象牙白的色调在柔和灯光的渲染下，为空间营造出家的温暖。餐厅是家的“胃”，半开放式的门，为欧式风格的空间增添了一份暖意。卧室的家具，采用经典元素，给人极具亲和力的人文感受，犹如著名的波尔多左岸盛产的红酒，香醇浓郁，悠远流长。简欧风格的书房，充满浓浓的文化气息，彰显了业主深厚的文化底蕴。

Mountain Top Residence

山顶大宅

◎ Design Company: Danny Cheng Interiors Ltd.
◎ Designer: Danny Cheng
◎ Location: Hong Kong, China
◎ Area: 666 m^2
◎ Main Materials: Marble, Wood, Clear Glass, Snakeskin, Aluminum Windows, Patterned Wallpaper

◎ 设计公司：Danny Cheng Interiors Ltd.
◎ 设计师：郑炳坤
◎ 地点：中国香港
◎ 面积：666 m^2
◎ 主要材料：云石、木材、清玻璃、蛇皮、铝窗、花纹墙纸

The project is a four-level mountain top luxury villa full of personality and the beauty of the architecture. Every floor is characteristic and fluent in the whole space. The villa has an inherent advantage to see the endless sea view.

The space mainly colors in bright colors and the brown color, linking the space in every level, creating a warm housing breath, elegant and dignity. The designer creates a magnificent space highlighted by different materials such as marble, wood, clear glass, snakeskin, aluminum windows, patterned wallpaper. In the open space, the picture frames are magnified and unified to fullfil the house, accented with special art works, adding a touch of art in every corner of space.

Four adjustable angle screens casually divide the living room and the dining room, creating the sense of layering. The huge chanderlier in the diniing room echoes with the screens in a hierarchical design.

本案是一幢 4 层高的山顶别墅，充满个性与建筑美。层与层之间各有特色且不失连贯，空间设计统一中彰显个性。本案拥有先天的优势，可饱览一望无际的海景。

在客厅和餐厅之间有 4 块可调整角度的屏风，营造出层次感的同时，也在不经意间将客厅与餐厅分隔开来。餐厅的巨形水晶吊灯，极富层次感的设计，与屏风相呼应。

空间以鲜艳的色调及深浅不一的棕色为主，使层与层之间保持连贯，营造出温暖的家居感觉，高贵、优雅又不落俗套。设计师还利用不同物料，如云石、木材、清玻璃、蛇皮、铝窗、花纹墙纸等来突显空间格调与气派。宽敞的大空间里，放大的画框款式统一，在视觉上不会有大而空的感觉，再配以别致的艺术品，每一个角落都赏心悦目，绽放艺术的光彩。

The Lemon House

香港大宅

- Design Company: Danny Cheng Interiors Ltd.
- Designer: Danny Cheng
- Location: Hong Kong, China
- Area: 333 m^2
- Main Materials: Wood, Marble

- 设计公司：Danny Cheng Interiors Ltd.
- 设计师：郑炳坤
- 地点：中国香港
- 面积：333 m^2
- 主要材料：木材、云石

THE LEMON HOUSE

This two-storey apartment adopted sharp and fashionable yellowish color as its major tone, which goes well with the surrounding green and thick wood, and add to the home a kind of refreshing and energetic spirit. Being inside it makes one feels like on holiday. The renovated apartment takes the advantage of its original structure: an extraordinary great internal height, which brings forth a sense of exceptional spaciousness. The French window was intentionally retained so that the outdoor scenery can be drawn into the fullest extent, while at the same time let in sufficient daylight. What is more, the variation of the natural scenery day and night brings to the house a unique sentiment and atmosphere.

An irregular-shaped yellowish sofa was chosen for the big parlor, which is trendy and sizeable, adding to the parlor a sense of fashion while echoing with the overall yellow theme. The tailor-made carpet made of cow skin integrates well with the outdoor natural environment as well. A movable screen is utilized to separate the parlor and the study, resulting in a clear layout and a strong sense of transparency and greatly enhanced the overall spaciousness. The Puzzle sofa zone right beside the window makes good use of natural lights and allows one to appreciate the outside view while resting and gets completely relaxed. It is also a perfect area for reading. As to material use, the designer skillfully chooses some bright colors and cloth with various textures, creating a brilliant and vivid internal environment. The living room on the second floor sustains the yellow theme of the whole design, in which a colorful and clear-layered sofa, made of cloth, enriches the tranquil space with some nice hue. Together with the outdoor daylight, the whole area is brought to life in an instant. Simplicity is the major style for the master bedroom. With a huge comfortable bed and wooden flooring, a sense of harmony and coziness is fully expressed.

CHANEL
CHANEL
GAIA
GAIA

The design and layout of the entire house has brought forth a modern, refreshing as well as distinctive dwelling, allowing the owner to escape from the bustle and hustle of the city and enjoy a feeling of vacationing at home.

本案是一座两层高大宅，以时尚、鲜艳的黄色为主调，周边翠绿茂密的山林，为家居空间注入一股清新与活力，亦增添了些许度假气氛。大宅先天条件优越：大厅的楼底很高，拥有十足的空间感。特意保留的落地玻璃窗，可将窗外美景尽收眼底，也有充足的光线进入室内，而且日夜的更替及自然景观的变化，也为宅子带来不一样的感觉和气氛。

大厅选用不规则的黄色沙发，既为大厅增添了时尚感，又与黄色主题相呼应。用牛皮制作的地毯亦与户外的自然气息融为一体。活动屏风区分出客厅与书房，区域分明，通透性强，大大增加了透明度及空间感。靠窗放置的 Puzzle 沙发区域采用天然光之余，亦可于休息时欣赏窗外景致，放松心情，是供休憩、阅读的好地方。在材料的运用上，设计师巧妙地采用鲜艳的颜色及不同质感的布料，令室内环境鲜明、活泼。二楼的客厅继续以黄色为主调，用色彩丰富、层次感强的布沙发为宁静的空间增色不少，再配合户外的天然光线，空间瞬间生机勃勃。主人房以简洁为主，舒适的大床和木地板，营造出和谐、舒适的感觉。

全屋的设计与布局，营造出一种时尚、清新而独特的感觉，让业主远离紧张的都市，于家中便可感受到度假时的悠闲。

Abbots Way

Abbots Way 别墅

◎ Design Company: AR Design Studio
◎ Designer: Andy Ramus
◎ Photographer: Martin Gardner

◎ 设计公司：AR 设计工作室
◎ 设计师：Andy Ramus
◎ 摄影师：Martin Gardner

Abbots Way, the latest creation by AR Design Studio, is a stunning five bedroom house. Bordered by mature trees and a small lake, this spectacular house creates feelings of ultimate relaxation and privacy, whilst its contemporary design juxtaposes superbly with its beautifully rural location, on the south coast of England.

The site is accessed via a private forest lane. Its overtly linear approach extends into the plot and creates the conceptual main axis. These dividing walls create four separate garden experiences, the first of which being the entrance space. The second is a calm pebbled Japanese herb garden to the rear of the house linking the separate office. The third is a private enclosed wooded space dedicated to the kitchen and master bedroom above. The fourth is a large open expanse of tropical plants and lawn reaching down to the lake, reflecting the open plan living space that looks over it.

A white box, sat on the top of the axial walls, gives the upper sleeping floor a light weight image and appears to be floating above the stone axes. It contains the five bedrooms, all with large glazed areas. This provides beautiful tree top views of the lake and surrounding woodland, creating a sense of being nestled amongst the tree canopy whilst lying in bed.

The first floor is living areas, nestled under the box and formed by the axis, are enclosed with large sliding glazed panels that provide a seamless link between the internal and external. The terrace further accentuates this, with the use of a single style of floor tile inside, that extends outside through flush thresholds to really give a true sense of "inside and outside living".

Abbots Way 别墅是 AR 设计工作室最新设计的作品，内有 5 间卧室。别墅周围被茂密的树木和一个小型湖泊环绕，营造出一种极度舒适和隐蔽的居住氛围。其现代的造型和外观，与英国南海岸优美的乡村环境完美融合。

别墅外有一条木质的小径，经由此可直达别墅。庭院内的独立墙体将整个空间分成 4 个区域：第一个区域为别墅主入口；第二个区域为别墅后方的日式花园，其与独立书房相连；第三个区域为厨房和主卧上方的木质封闭空间；第四个区域为湖泊旁的热带植物群和草坪。

别墅主墙的上方有一个巨大的白色盒体，从远处望去，犹如漂浮在石墙上一般。在白色盒体内的房间里均可欣赏到优美的湖光及周围的丛林，休憩于此，犹如栖于枝头，飘飘欲仙。

别墅的一楼是起居空间，居于盒体的下方，巨大的玻璃滑动门板打破了室内外的界限。向户外延伸的阳台，与室内处于同一水平线，且采用与室内风格相同的地砖，真正实现了室内外生活的完美过渡。

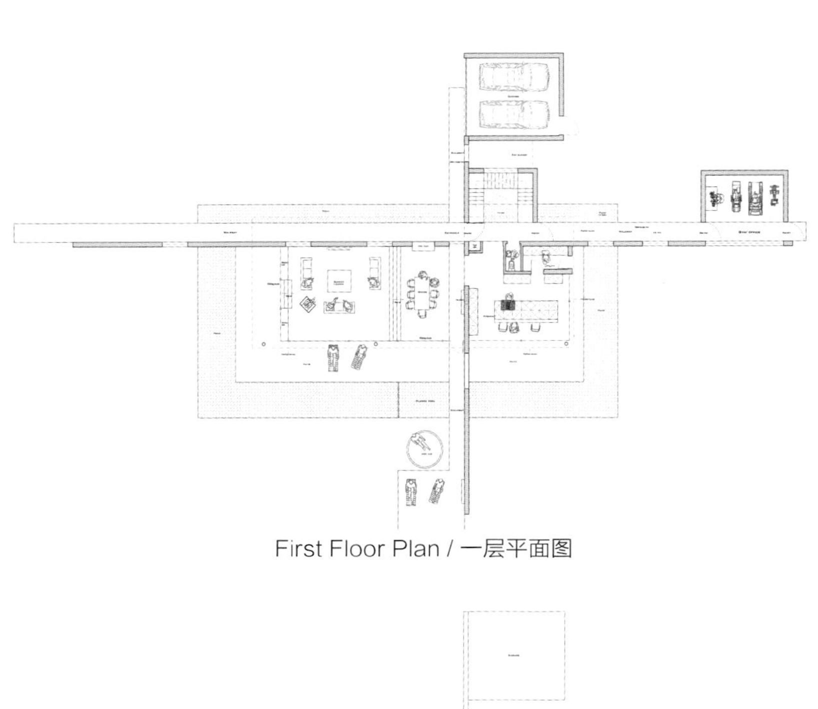

First Floor Plan / 一层平面图

Second Floor Plan / 二层平面图

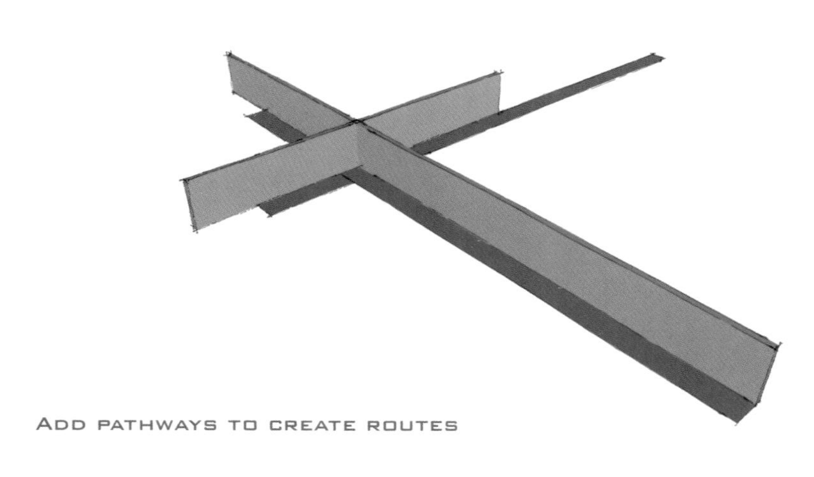

ADD PATHWAYS TO CREATE ROUTES

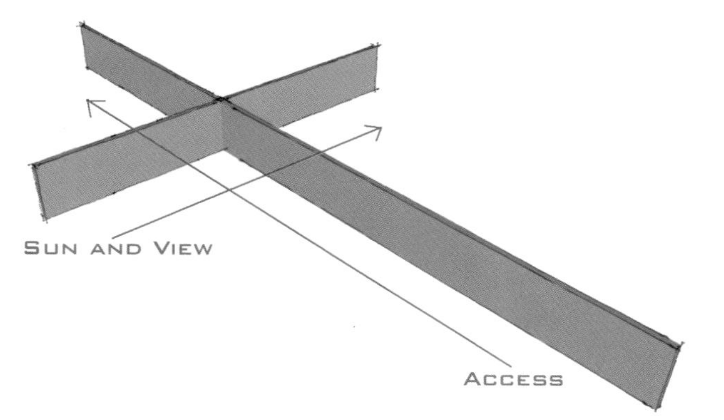

BUILD WALLS TO ALIGN WITH SITE AXES

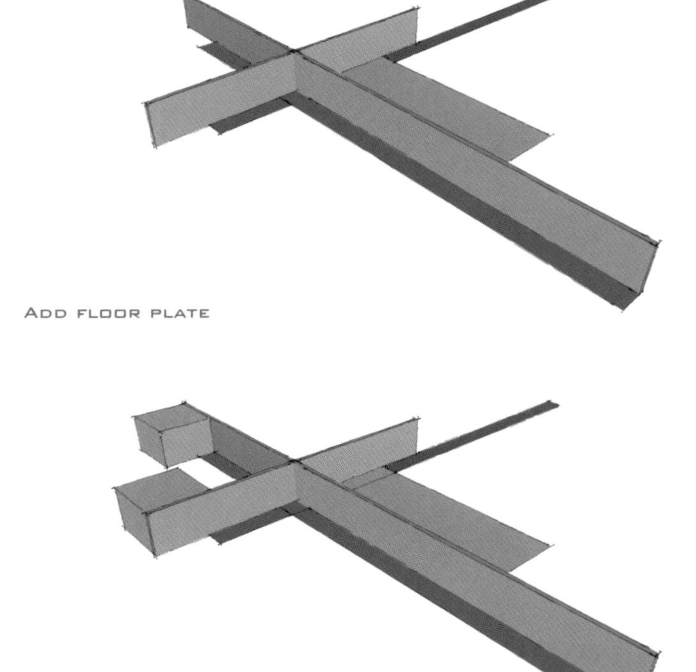

ADD FLOOR PLATE

ADD BLOCKS FOR GARAGE AND GYM

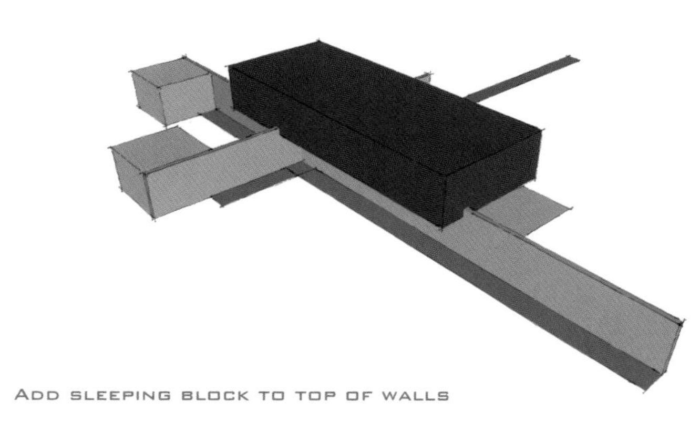

ADD SLEEPING BLOCK TO TOP OF WALLS

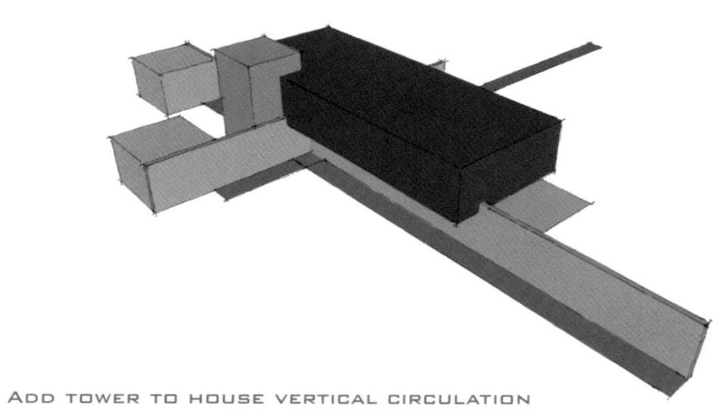

ADD TOWER TO HOUSE VERTICAL CIRCULATION

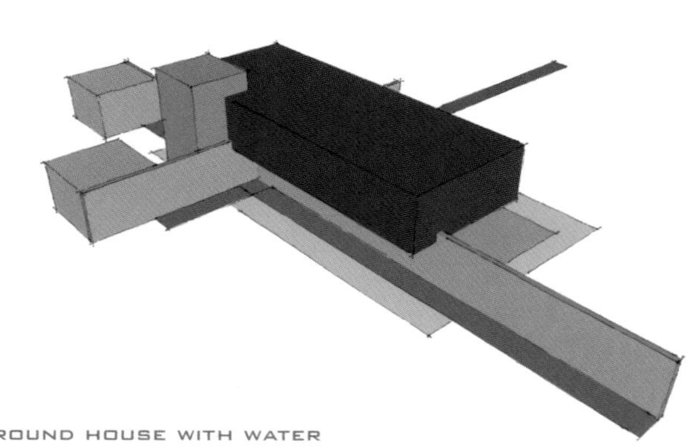

SURROUND HOUSE WITH WATER

The Lighthouse 65

灯塔 65

◎ Design Company: AR Design Studio
◎ Designer: Andy Ramus
◎ Photographer: Martin Gardner
◎ Location: Hampshire, UK
◎ Area: 180 m^2

◎ 设计公司：AR 设计工作室
◎ 设计师：Andy Ramus
◎ 摄影师：Martin Gardner
◎ 地点：英国汉普郡
◎ 面积：180 m^2

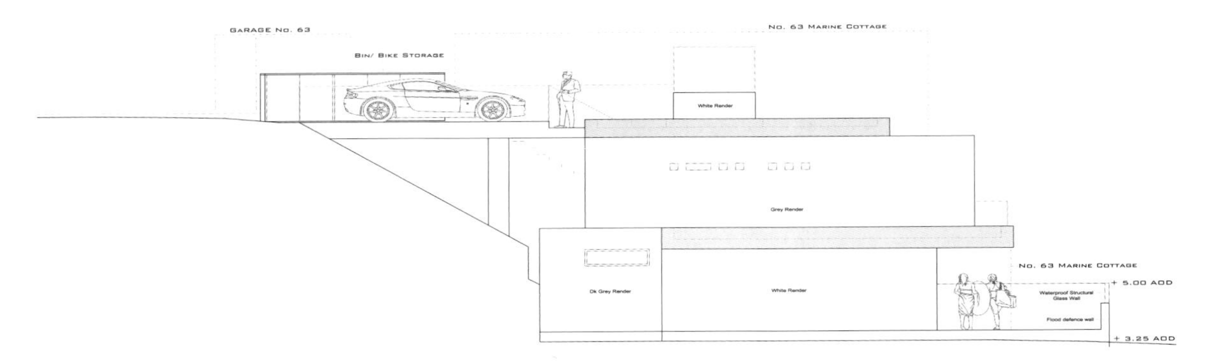

West Elevation / 西立面

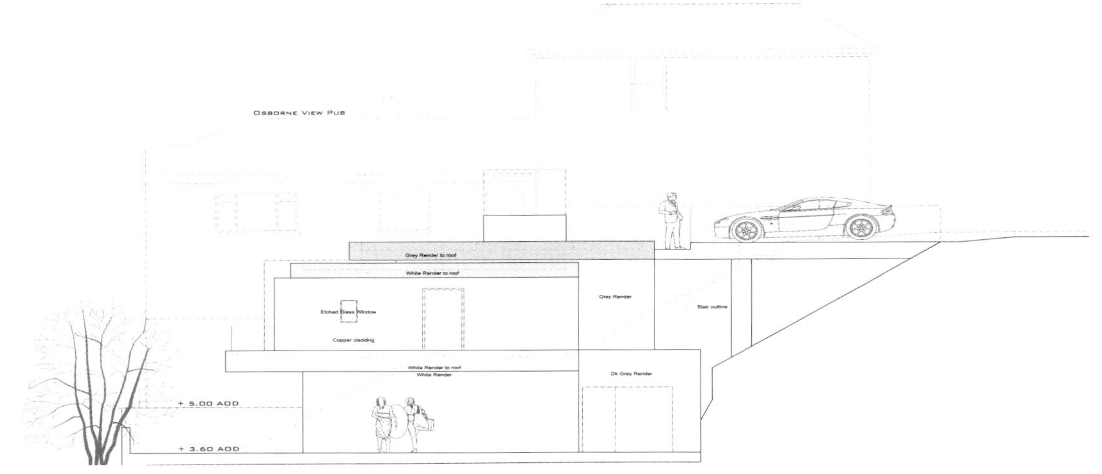

East Elevation / 东立面

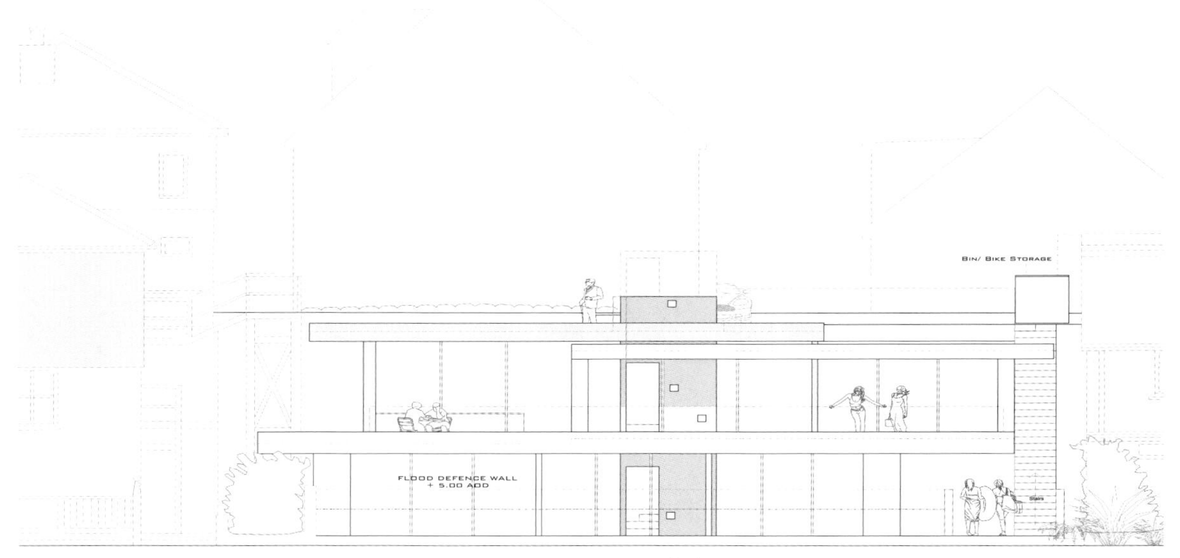

South Elevation / 南立面

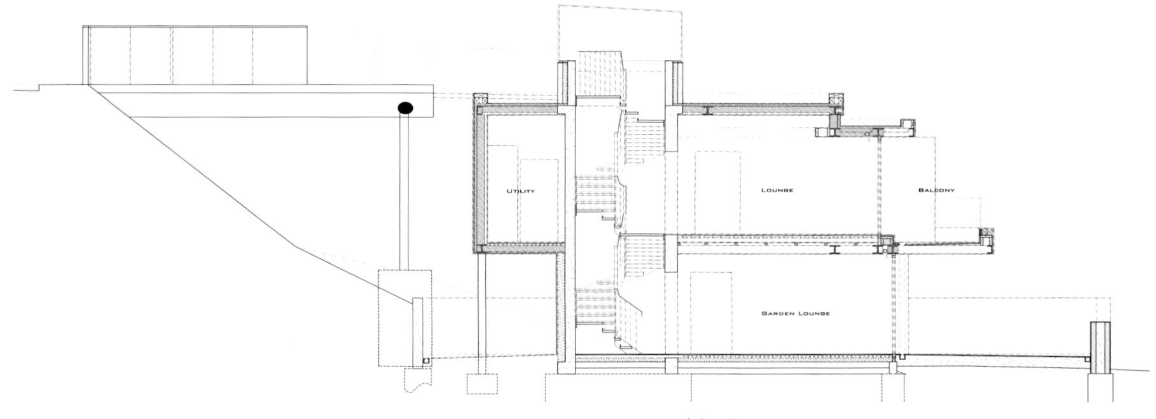

The Section Drawing / 剖面图

The Lighthouse 65 is a beachfront property on the south coast of England. It is a super insulated, luxury 3 bedroom house sitting in a beautiful water side location enjoying stunning views of the Solent and the Isle of Wight.

The site is enclosed between two neighbouring buildings and a 7 meters high embankment to the north; of which pavement and street access sits at the top. This access, to the roof level of the property, and the one-directional view over the beach, English Channel and Isle of Wight to the south.

The design concept was to maximize the building's width, so every key room enjoys expansive views of the vista. All bathrooms and utility spaces run at the rear of the property, allowing the view to be continuous for all living spaces. The house sits 7 metres below road level with the roof acting as a parking deck for 3 cars. Visually the roof and floor decks are hung from the central concrete core, terminating in large cantilevers that provide shade and open-air shelter to the ground floor.

Balconies and outdoor accommodation are provided by horizontal planes carefully cantilevered from a central access tower which penetrates the roof plane; atop of which sits a frameless glass enclosure providing access and acting as a lighthouse. Lighting within this enclosure is triggered by a barometer providing instantaneous information to the beach and sea beyond. At night the lit glass glows to indicate local weather conditions.

灯塔 65 是英国南部海岸的一座海滨别墅，其为一座超保温的豪华三居室，坐拥迷人的海景，享受着索伦特和怀特岛的壮丽景观。

灯塔 65 被两个相邻的建筑包围，北面有一个 7 米高的筑堤，人行道和街道入口位于别墅的顶部。别墅顶层的入口，高于海滩，可远眺英吉利海峡和南部怀特岛的动人景色。

本案设计师 Andy Ramus 最大限度地提升了建筑物的高度，使每一个主要房间都可以享受到远处广阔的景色。所有的浴室和实用空间均被安排在房屋的后面，让所有的生活空间构成一个连续的视野。房子坐落在低于街道 7 米的地方，屋顶可作为 3 辆汽车的停车场。在视觉上，屋顶和地面甲板都悬挂于中心混凝土的核心，终点位于为地面提供阴凉和露天庇护的大悬臂梁上。

阳台和露天住宿区由悬空的水平面支撑，由无框玻璃外壳覆盖的塔穿过屋顶平面，为建筑提供入口的同时，也可作为灯塔。玻璃外壳内的照明是一个为海滩和大海以外的地方提供瞬时信息的晴雨表，以此来显示当地的天气状况。

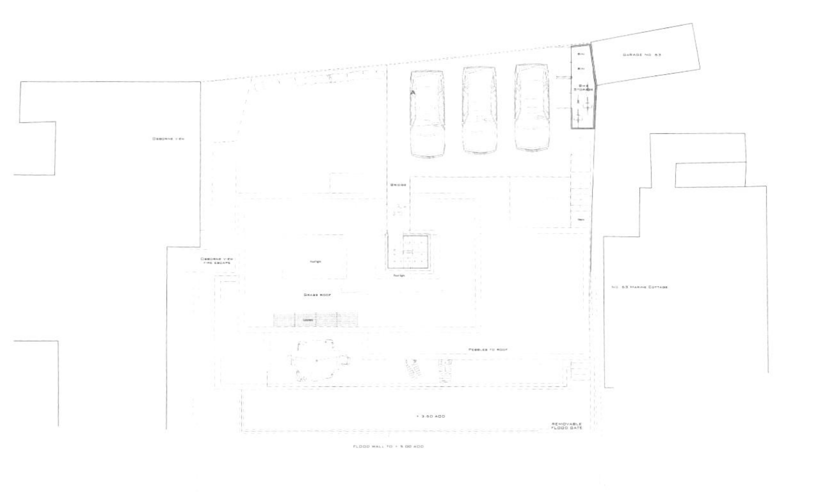

Roof Plan / 屋顶平面图

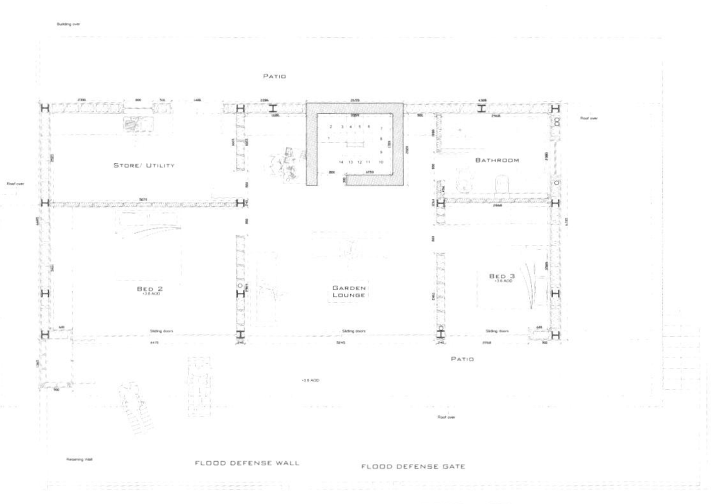

First Floor Plan / 一层平面图

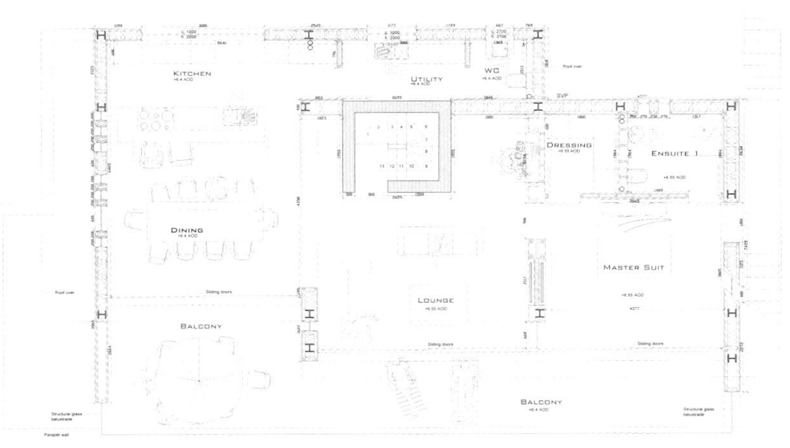

Second Floor Plan / 二层平面图

Joc House

Joc 别墅

◎ Design Company: Nico van der Meulen Architects
◎ Designer: Rudolph van der Meulen

◎ 设计公司：Nico van der Meulen Architects
◎ 设计师：Rudolph van der Meulen

Richard Rogers

Richard Rogers

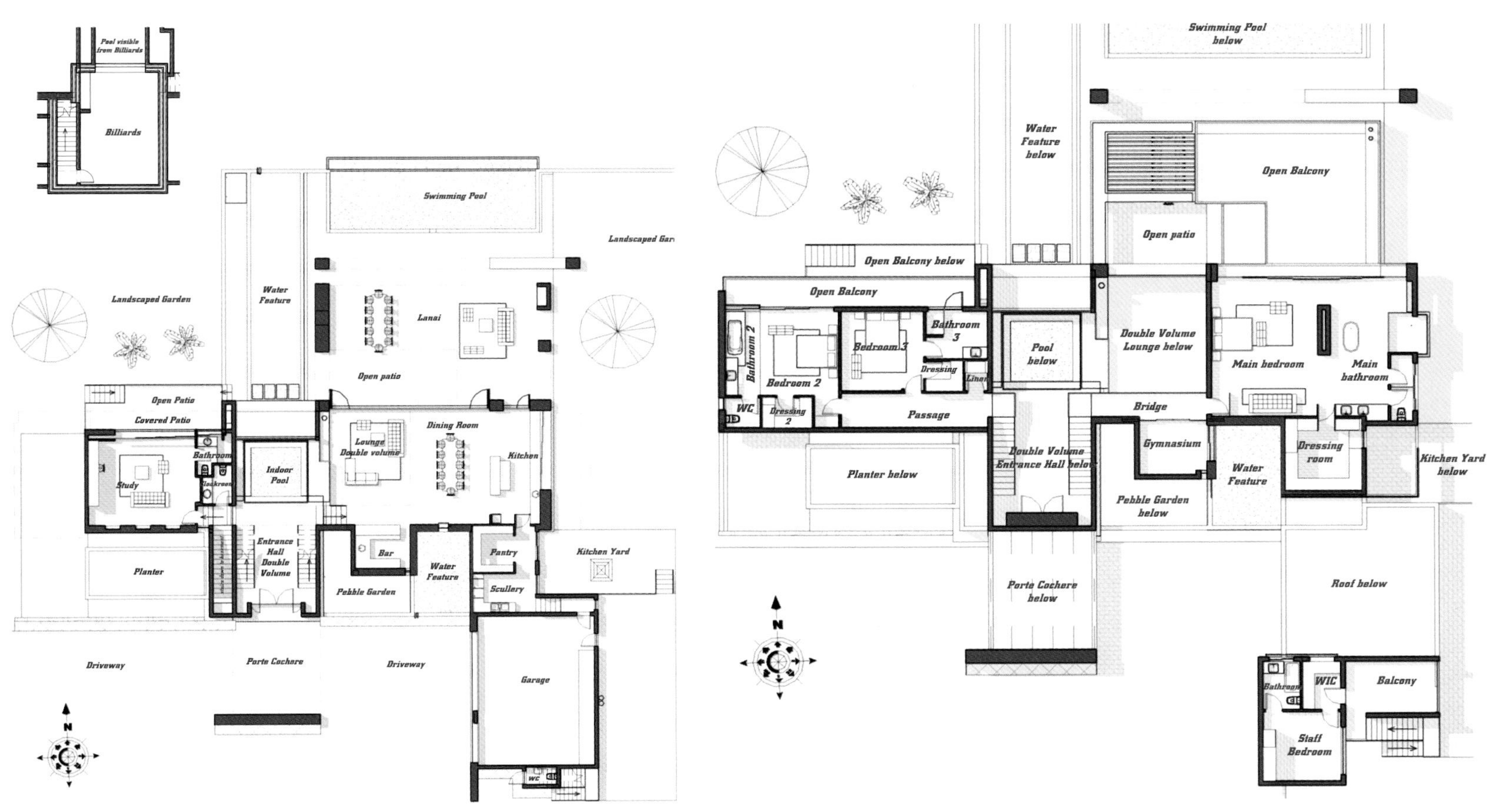

First Floor Plan / 一层平面图

Second Floor Plan / 二层平面图

This villa was designed by Rudolph van der Meulen of Nico van der Meulen Architects for a mature couple on a 10,000 square meters parcel of land on a northern slope overlooking a private game reserve. The total covered area is nearly 900 square meters. The driveway takes you past the house to the back to a protected porte cochere and a glass front door, from where you overlook an indoor splash pool and an outdoor water feature with a series of fountains spilling over the far edge.

The entrance hall overlooks a double volume, open plan dining room and kitchen leading through frameless glass doors onto an expansive lanai and infinity edge pool, framed by a massive beam. The wall between the family room and lanai is all glass and can be fully opened. The kitchen is directly connected to the lanai, with the barbeque within 3 meters from the kitchen counter. Behind the kitchen are a walk-in pantry, cold room, scullery and laundry with a connection to the garages.

The double volume living area has high level windows above the roof of the lanai, letting the light in summer and sun in winter. Due to the fact that large areas of glass can be opened to the cooling breezes from the pool and water feature, air conditioning was not necessary. On the one side a main bedroom and bathroom, divided by a low wall with a double sided fireplace. The basins are installed in this wall. The other wing consists of a study and guest bedroom downstairs, and 3 kid's suites upstairs. All the habitable rooms in the house are north facing, with solar control carefully considered resulting in a cool house in summer and a bright, sunny and warm house in winter, with a minimum of heating and cooling required.

Joc别墅由Rudolph van der Meulen设计。业主是一对中年夫妇。本案占地面积900平方米，于北部的斜坡处可远眺野生动物保护区。沿车道前行，经由别墅可直达后门，沿途可俯瞰室内泳池、户外水景及从远处悬崖倾泻而下的瀑布。

入口大厅拥有超大的体量，内设家庭活动室、餐厅和厨房，家庭活动室和阳台之间采用无框玻璃门，可完全敞开，整合空间。厨房与阳台相连，阳台上的烧烤架距厨房柜台不足3米，厨房的后面是茶水间、冷藏室、后厨房和与车库相连的洗衣房。

沿着楼梯向上走，则进入了二层，双高的客厅位于一层阳台的上方，大面积的落地窗，为室内引入更多光线的同时，也将户外微风和水景引入室内，于此可享受夏日的凉爽。主卧和主卫位于一侧，中间被一道矮墙区隔开来，矮墙两面都饰有壁炉，朝向主卫的墙体上安装了盥洗池。空间的另一侧，设有书房、客卧和儿童房，别墅内的所有房间均朝北，这样可以很好地控制光照，使室内冬暖夏凉，最大限度地降低能耗。

House Refit

Refit 别墅

◎ Design Company: TG Studio
◎ Designer: Thomas Griem
◎ Photographer: Philip Vile

◎ 设计公司：TG 工作室
◎ 设计师：Thomas Griem
◎ 摄影师：Philip Vile

This modern terraced house on a private estate in North London has been completelytransformed - from dark and soulless into an oasis of cool, calm and contemporary. Thomas Griem, Design Director of TG Studio, with architecture and design offices in London, has designed a beautiful home by entirely re-organising the internal layout to let natural light flow through every corner of the house.

Thomas tackled the brief to create a light and bright space and make the most of the unusual layout by designing a new central staircase, which links the six half-levels of the building. A minimalist design with glass balustrades and pale wood treads connects the upper three floors consisting of three bedrooms and two bathrooms with the lower floors dedicated to living, cooking and dining. The staircase was designed as a focal point, one you see from every room in the house. It's clean, angular lines add a sculptural element, set off by the minimalist interior

Anish Kapoor
ISABELLA

of the house.

The owner's Scandinavian roots are echoed in every aspect of the design with exclusively white walls and ceilings and white pigmented pine floors as a backdrop to furniture, furnishings and accessories in natural materials such as linen, pale wood, glass and stone, all in white and neutral colours.

Refit 别墅坐落于伦敦北部的私人地块上，设计完成后本案焕然一新，已然成了一个冷静、低调和现代的建筑。TG 工作室的设计总监 Thomas Griem 经过全面的规划和布局后，力求打造出一个完美的家居空间，让自然光线洒入空间的各个角落。

在这个超乎寻常的空间内，设计师 Thomas Griem 通过新增的中央楼梯来连接内部各个空间，力求营造一个宽敞、明亮的内部空间。建筑上部的三层空间包含 3 个卧室和 2 个浴室，其楼梯由玻璃围栏和浅木色踏板组成。下层空间为公共区域，包含起居室、厨房和餐厅。楼梯处于建筑的核心，在室内的任何一个房间都可以看到，其简洁的弧线与空间极简主义风格相吻合。

本案极简主义的设计理念已深深根植于 Refit 别墅主人的心中，空间中白色的墙壁、天花、浅色的地板与室内的家具、配饰和天然的材质相呼应，室内的一切都呈现出白色和中性的色调。

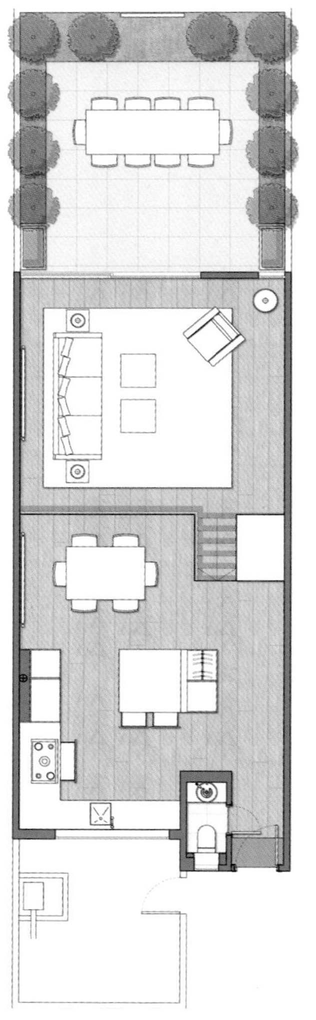
First Floor Plan / 一层平面图

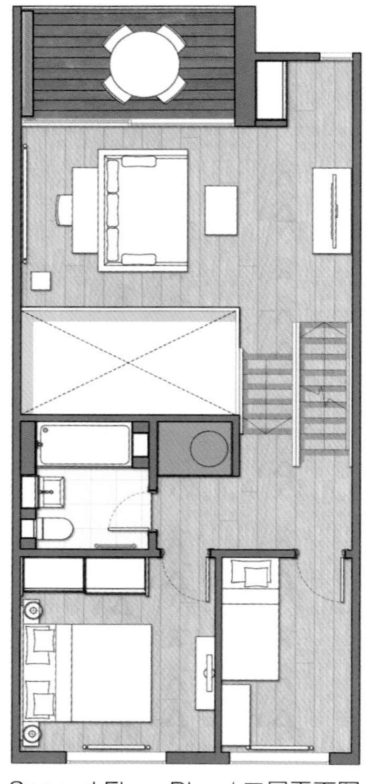
Second Floor Plan / 二层平面图

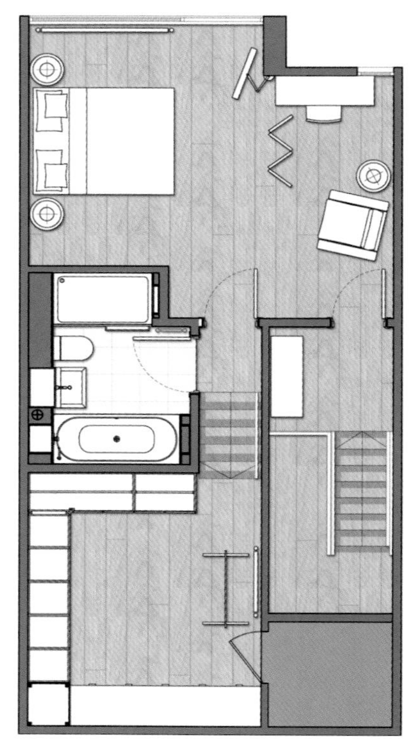
Third Floor Plan / 三层平面图

19
MON IX

Apple Bay House

苹果湾别墅

◎ Design Company: Parsonson Architects Ltd.
◎ Designer: Gerald Parsonson, Craig Burt, Daniel Watt
◎ Photographer: Paul Mcredie
◎ Location: Wellington, New Zealand

◎ 设计公司：Parsonson Architects Ltd.
◎ 设计师：Gerald Parsonson、Craig Burt、Daniel Watt
◎ 摄影师：Paul Mcredie
◎ 地点：新西兰惠灵顿

The property is located in the Marlborough Sounds, which is at the north end of the South Island of New Zealand. It is accessible by 4 wheel drive vehicle and boat from the nearest township Picton.

It is a holiday house for extended family and sits on a west facing bush clad hillside in Apple Bay. The designer intended to create a delicate platform amongst trees for life to unfold. The house is in two parts, each with a fine roof that follows the slope of the land. The upstairs living areas are configured as a large elongated space with decks at each end to enjoy both the morning and afternoon sun and with retreat and service spaces serving these. Downstairs bedrooms float amongst the trees with windows opening across the treetops. The gap and junction between the two building parts is a continuation of the path up from the boat house.

苹果湾别墅位于新西兰南岛北端的马尔堡峡湾，临近车船往来的港口城市——皮克顿。

这是一个大家庭的度假别墅，西部朝向灌木丛生的苹果湾山脉。设计师试图在丛林密布的山中建造一个平台，供生活、休憩之用。整个建筑共包含两个部分，别墅依山就势而建，上层是起居空间，客厅的每一侧都设置了甲板，主人可在此欣赏到美丽的晨曦和落日。下层的卧室空间似漂浮在树上一般，打开窗子即可看到从中穿过的树梢。一条从船屋延伸而出的小径将两个建筑连接起来。

Hall
Bed 1
Bath 1
Laundry
Bed 2
Bunkroom
Bath 2
Bed 3
Hall
outdoor shower
Link
store
WC
Store
Deck
Dining
Kitchen
Snooker
Living
Deck
Entry
Stage / Library
Pantry
WC
Store
Utility
Car Deck

Seaview House

海景别墅

◎ Design Company: Parsonson Architects Ltd.
◎ Photographer: Paula Mcredie
◎ Location: Wellington, New Zealand

◎ 设计公司：Parsonson Architects Ltd.
◎ 摄影师：Paul Mcredie
◎ 地点：新西兰惠灵顿

The property is located in Wellington, New Zealand. It sits just below the road facing east, looking over the botanical gardens and Wellington Harbour. A neighbouring house sits much higher to the north with another lower to the south. The site loses sun directly to the north, but receives both generous morning and afternoon sun.

The owners have a large family, with both older and younger children. The house is arranged to accommodate these different age groups with bedrooms on different levels and a variety of living spaces in the middle with walls to house art and a swimming pool for the keen swimmers in the family.

Simple corrugated iron roofs wrap and frame the house, which help create a relationship with the houses of the area. The green color of these also helps the house recede into the backdrop of greenery. In contrast to this, and housing the garage and bedrooms, a more organic wooden clad element runs between the corrugated iron roofs.

海景别墅位于新西兰惠灵顿，别墅位于道路的下方，面东而建，可远眺植物园和惠灵顿港口。邻近的别墅高低不一，错落有致。别墅的朝向和位置虽然背光，但却收获了日出和日落的美景。

本案是为一个大家庭而建的，家庭成员分为 3 个不同的年龄段，父母和年幼孩子的卧室位于楼上，年长孩子的卧室和一个游泳池位于住宅底部，各种不同的休闲空间位于居室中部，可以俯瞰美景，享受阳光。

别墅外部的斜瓦楞铁皮屋顶，与周围的别墅在外貌上相匹配，建筑外部的绿色于周围的丛林中隐退了自己，并与车库、卧室上方的轻质木复合屋顶形成对比。

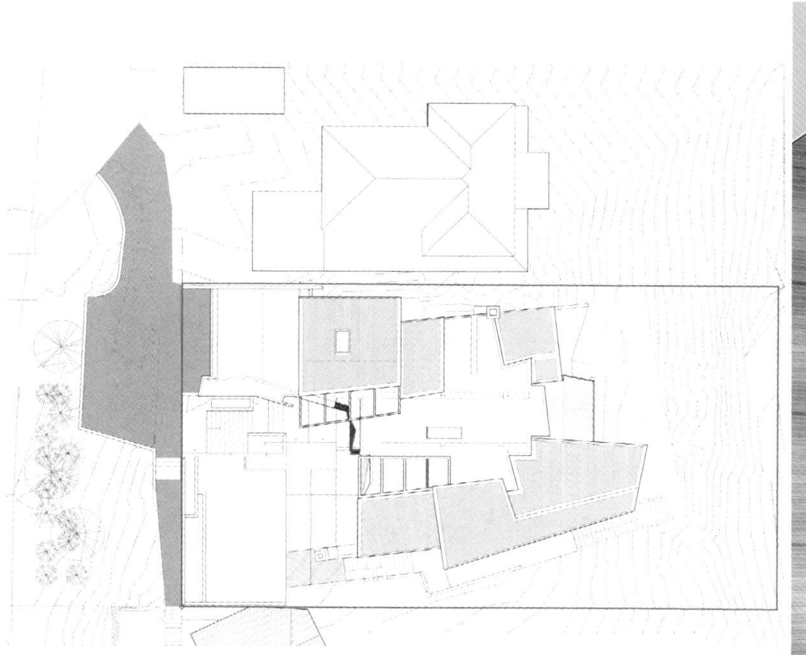

Site Plan / 平面图

Cultural Holiday Villa

人文度假别墅

- Design Company: Etai. Space Design Office
- Designer: Zhang Xianghao
- Area: 330 m^2
- Main Materials: Stone, Wood, Grey Glass, Black Mirror, Weathering Steel, Stainless Steel

- 设计公司：伊太空间设计事务所
- 设计师：张祥镐
- 面积：330 m^2
- 主要材料：石材、木材、灰玻璃、黑镜、耐候钢、不锈钢

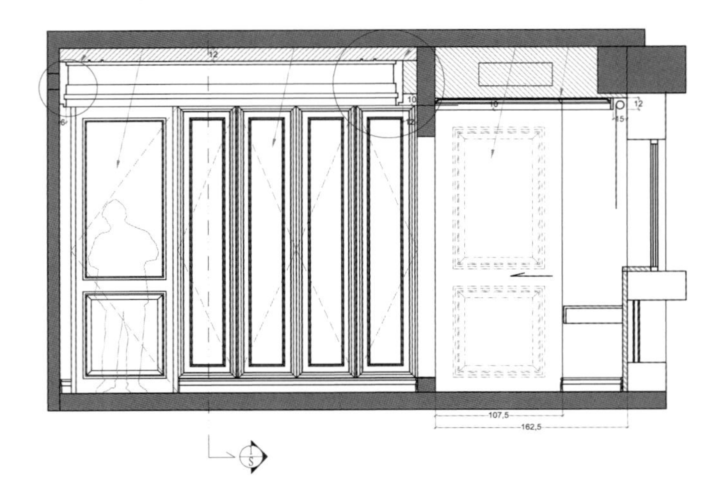

107,5
162,5

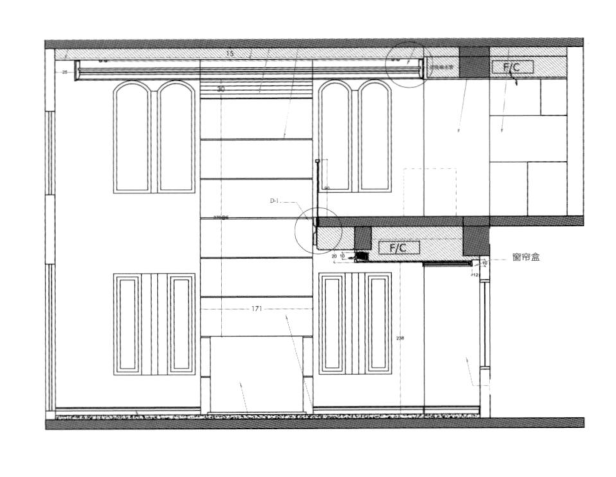

F/C
F/C
窗帘盒

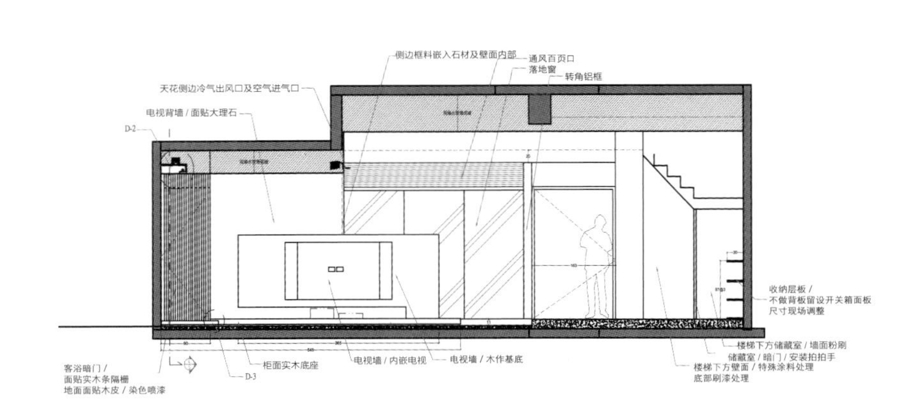

侧边框料嵌入石材及壁面内部
通风百页口
落地窗
转角铝框
天花侧边冷气出风口及空气进气口
电视背墙 / 面贴大理石
D-2
收纳层板 /
不做背板留设开关箱面板
尺寸现场调整
楼梯下方储藏室 / 墙面粉刷
储藏室 / 暗门 / 安装拍拍手
楼梯下方壁面 / 特殊涂料处理
底部刷漆处理
客浴暗门 /
面贴实木条隔栅
地面面贴木皮 / 染色喷漆
柜面实木底座
D-3
电视墙 / 内嵌电视
电视墙 / 木作基底

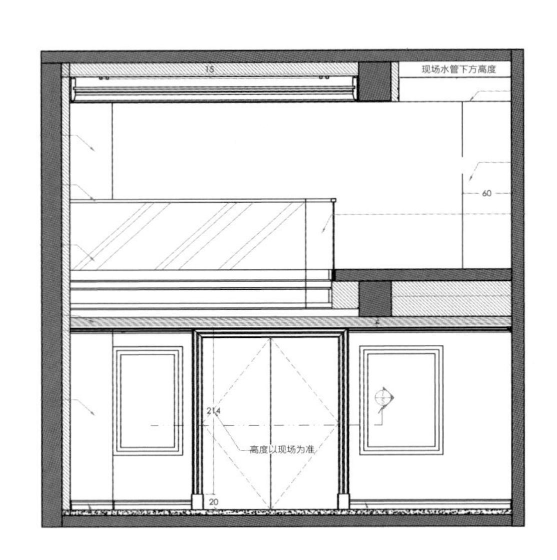

现场水管下方高度
高度以现场为准

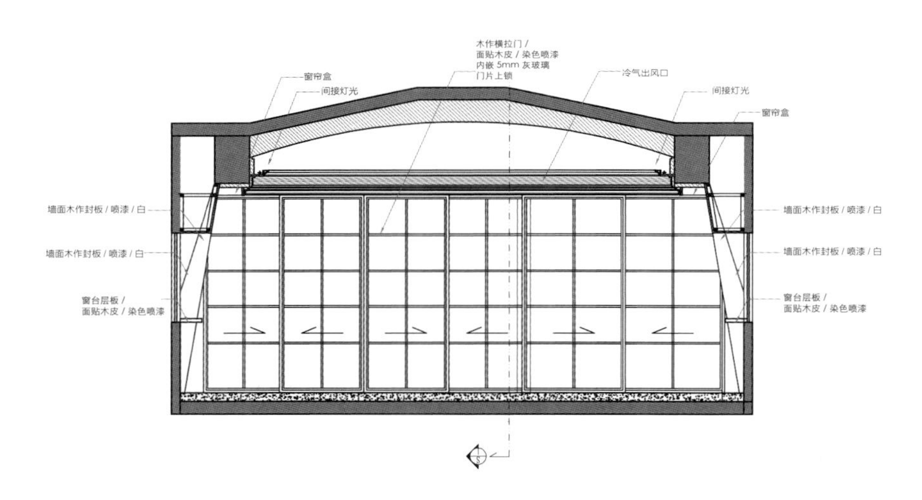

木作横拉门 /
面贴木皮 / 染色喷漆
内嵌 5mm 灰玻璃
门片上锁
窗帘盒
间接灯光
冷气出风口
间接灯光
窗帘盒
墙面木作封板 / 喷漆 / 白
墙面木作封板 / 喷漆 / 白
窗台层板 /
面贴木皮 / 染色喷漆
墙面木作封板 / 喷漆 / 白
墙面木作封板 / 喷漆 / 白
窗台层板 /
面贴木皮 / 染色喷漆

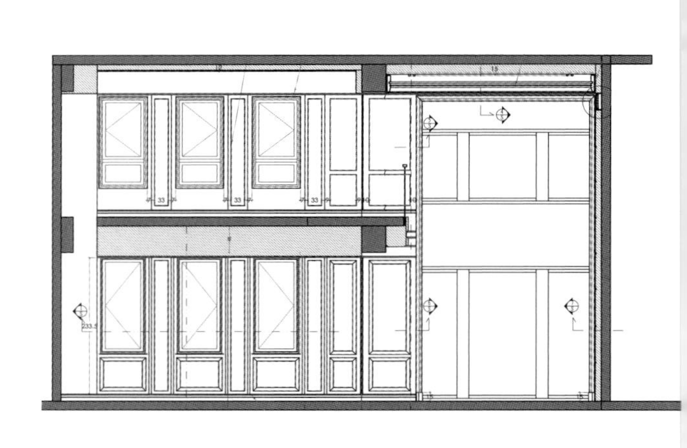

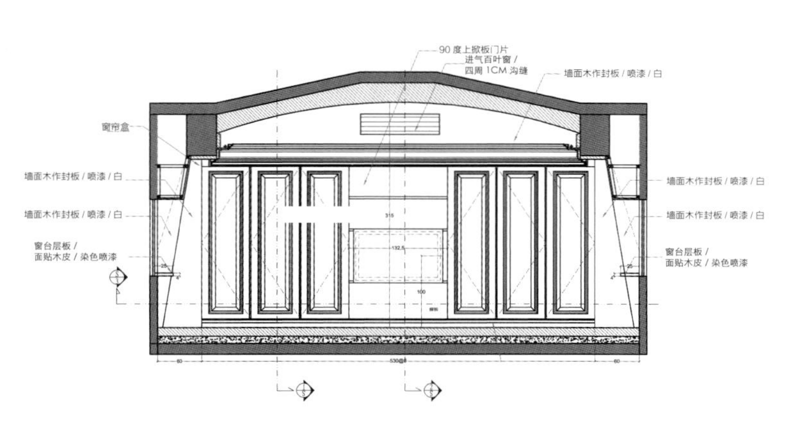

90 度上掀板门片
进气百叶窗 /
四周 1CM 沟缝
墙面木作封板 / 喷漆 / 白
窗帘盒
墙面木作封板 / 喷漆 / 白
墙面木作封板 / 喷漆 / 白
窗台层板 /
面贴木皮 / 染色喷漆
墙面木作封板 / 喷漆 / 白
墙面木作封板 / 喷漆 / 白
窗台层板 /
面贴木皮 / 染色喷漆

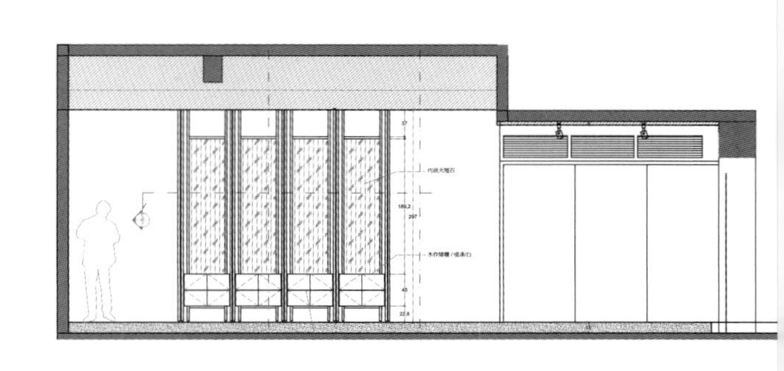

This project is located in Taipei, by which the land agent wants to deliver a life concept that keep away from the turmoil in the city to live a leisurely life. Therefore, the designer merges the understated and elegant elements into the project to creat a cultural holiday villa.

The cultural tone of the whole space is set from the parking area in the entrance. The sedate space in the first floor make resident comfortable and the light classic colors in the second floor please people.

To match with the overall tone, the colors in the space give priority to Beige, gray, brown, black and the original wood color to create a magnificent and sedate space. The soft installations is outstanding, which are collected by the designers, such as the carpentry in Thailand, the pottery and porcelain in Japan, the old artistic items in Philippines, creating a magnificent and wonderful space.

本案位于台北市，地产商意欲通过此样板间的设计传达出远离城市纷扰，享受悠闲居家生活的理念。因此，设计师将沉稳、低调与优雅融入本案，力求打造出一栋具有人文气质的度假别墅。

设计师从入口停车区开始，就奠定了整个空间的人文基调。一楼空间的沉稳，让居住者仿佛有了一种回归舒适居家生活的感觉。二楼明亮的轻人文古典色系，更让人心情沉静。

为配合空间的整体基调，空间以米色、灰色、褐色、黑色、原木色为主色，使空间大气而沉稳。本案的软装陈设也尤为考究，其陈设物件均为设计师之珍藏，泰国的木器、日本的陶瓷器、菲律宾艺术家的老物件等等，使空间大气、美观。

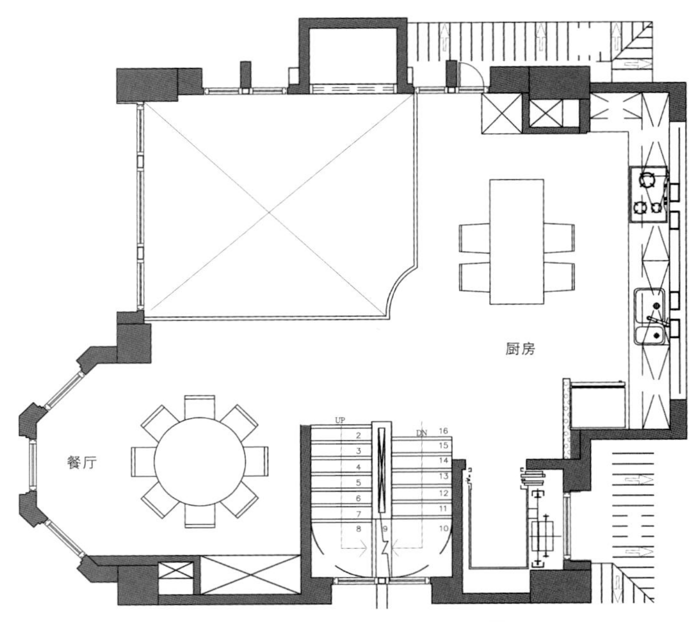

Third Floor Plan / 三层平面图

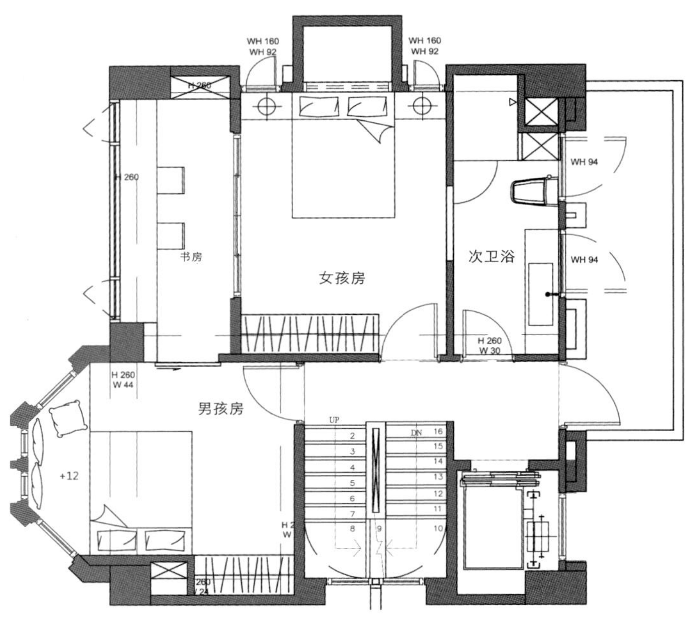

Fourth Floor Plan / 四层平面图

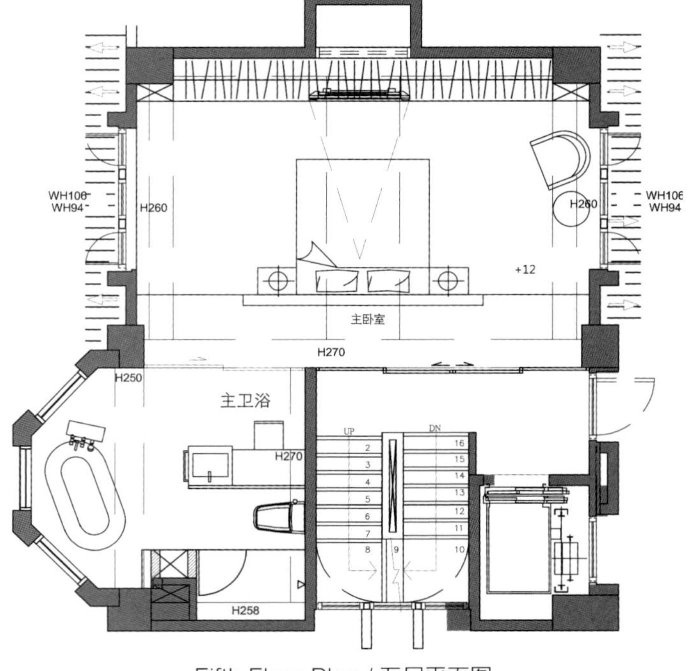

Fifth Floor Plan / 五层平面图

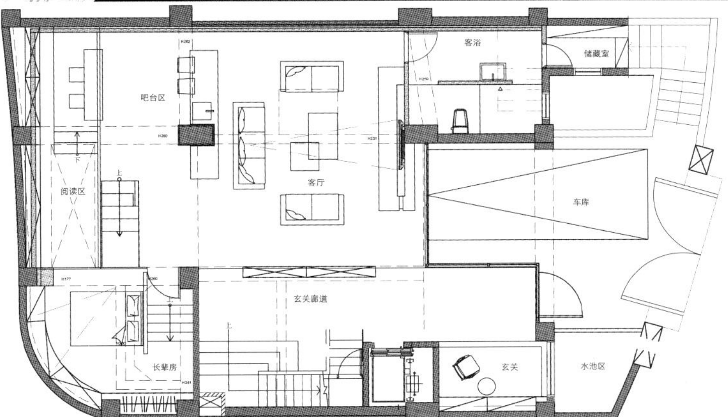

First Floor Plan / 一层平面图

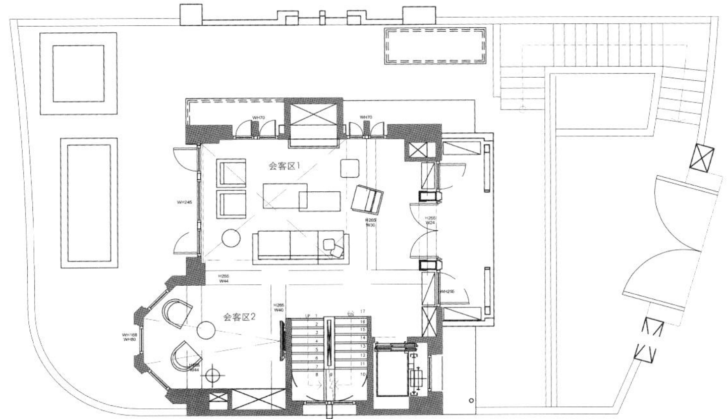

Second Floor Plan / 二层平面图

Taihu Tianque, Suzhou

苏州太湖天阙

◎ Design Company: Zhumu Design
◎ Designer: Chen Jie
◎ Area: 660 m^2
◎ Main Materials: Wooden Floor, Marble, Antique Brick, Tile, Mosaic, New Zealand Wool Carpet, Oak Solid Wood

◎ 设计公司：上海筑木空间设计装饰有限公司
◎ 设计师：陈洁
◎ 面积：660 m^2
◎ 主要材料：木地板、大理石、仿古砖、花砖、马赛克、新西兰羊毛地毯、橡木实木

Suzhou Taihu Tianque houses are designed in steel structure by German designers. The outskirt of the residences are Chinese style garden, matched with the inside European country style to form a strong impulse force.

The interior design continues the difference to create "fusion" to the extreme.

Heavy and Light

The project remains the original part to the extreme in a European country style. The living room skillfully uses soft decorations such as bamboo curtain, fabrics, floriculture to soften the steel structure to warm the whole space.

Cold and Warm

The steel structure shows an open and wide space in a European style. At the same time, considering the disadvantage of the house, a big seperated fireplace is set in the living room, ensuring the warmth in the winter and at home.

Privacy and Openness

The main bedroom maintains the slope crest and skylight, boldly

connecting with the bathroom. The warm decorations in the bathroom cover the cold design to continue the living space to set free the resident's soul.

Unreality and Reality

The underground level is heavy and calm. The designer uses the dark wood to act as hang ceiling, the stone as wainscot board, creating a real cellar full of mystery and interest.

苏州太湖天阙由德国设计师设计，内部均为钢结构。别墅外围皆为中式园林，与其内部的欧式乡村风格不期而遇，让中西建筑的迥异风格在这里碰撞出火花。

本案的室内设计将“迥”延续下去，打造出最高境界的“融”合。

重与轻

设计最大限度地保留了房屋的原始结构，沿袭欧式乡村风格。客厅巧妙利用竹帘、布艺、花艺等装饰，配以轻灵的设计手法，将原本凝重、刚硬的钢结构房屋柔化，使整个空间充满温情。

冷与暖

钢结构的房屋，空间高大、空灵，尽展粗犷、开放的欧式风情。同时，考虑到钢结构房屋节能的缺陷，设计师在客厅设置了大型的独立壁炉，除满足冬季取暖的需求外，也将“家”专属的温暖氛围烘托出来，两全其美。

私密与开放

主卧保留了原有的坡顶及天窗，并大胆地将卧室与卫生间完全打通，卫生间温馨、细腻的设计装饰掩盖了石材冰冷的质感，让居住者的心灵获得最大的自由。

虚幻与真实

地下室有地窖的感觉，厚重、沉稳。设计师采用深色的木质做吊顶，粗犷的石材做护墙板，营造出最真实的地窖世界，神秘中颇具趣味。

隆重出品
築木
空間
Summer Holiday

PUBLIC TELEPHONE
5
Chianti

Clam
Bake
LONG ISLAND FISHING PIER
E. NORTHPORT, NEW YORK
EST. 1933
Fresh
SEAFOOD
PUBLIC TELEPHONE
5¢
Chianti

PUBLIC TELEPHONE
5¢
Chianti

CATHY

Huizhou Golf Villa

惠州高尔夫别墅

◎ Design Company: KSL DESIGN（HK）LTD.
◎ Designer: Aday Lam, Wen Xuwu, Ma Huize
◎ Location: Huizhou, Guangdong
◎ Area: 2,000 m^2
◎ Main Materials: Camphor, Stone, Lymar Limestone, Leather, White Brushing Lacquer, Grey Wood Marble, Black Steel, Special Glass, Stone Mosaic

◎ 设计公司：KSL 设计事务所
◎ 设计师：林冠成、温旭武、马海泽
◎ 地点：广东惠州
◎ 面积：2 000 m^2
◎ 主要材料：樟木、秀石、白砂石、皮革、白色手扫漆、灰木纹大理石、黑钢、特殊玻璃、石材马赛克

Besides the luxury facade, the project also possesses the inner beauty. The space is highlighted by the elegant chanderlier and every luxury furniture, magnificent and fantastic. The designer reasonably creates a natural shape and a unified color to design the whole space which is transparent, bright, clear and magnificent. Get rid of the tedious part, the natural space provides more life imagination to the owner.

惠州高尔夫别墅除了拥有奢华的外表，还拥有“内在美”。高雅的水晶灯彰显出空间的大气，每一件奢华的家具都散发着空间内在的魅力。设计师合理利用整体空间，让造型自然舒适，色调和谐统一。唯美硬朗的线条与精心挑选的大理石等材料自然融合，使空间显得通透明亮、素雅利索而又大气稳重。在本案空间中，设计师摒弃了别墅空间一惯的繁琐与厚重，亲切自然而又张弛有度的空间为业主的生活提供了更多想象的空间。

228

Winds to Aegean Sea

风向爱情海

◎ Designer: Tao Sheng
◎ Photographer: Tao Sheng
◎ Location: Nanjing, Jiangsu
◎ Area: 200 m^2

◎ 设计师：陶胜
◎ 摄影师：陶胜
◎ 地点：江苏南京
◎ 面积：200 m^2

The project creats a Mediterranean atmosphere, adding a touch of nature and ocean. We always enjoy a cozy life when we keep away from the outside and back to home sitting in a comfortable sofa. We may occasionally feel the warmth of the home where we can find what we need.

本案营造出一种地中海的氛围，其自然气息扑面而来，更有一种安逸、舒适的海洋气息。每当我们拂去外界的浮尘，回到家里安逸地躺在沙发上，这种惬意就会油然而生，令我们乐此不疲。一个家，一种寄托，也许在不经意间我们已然感受到家的美好。

Our House Rules
Relax. Relax. Relax.
LOVE

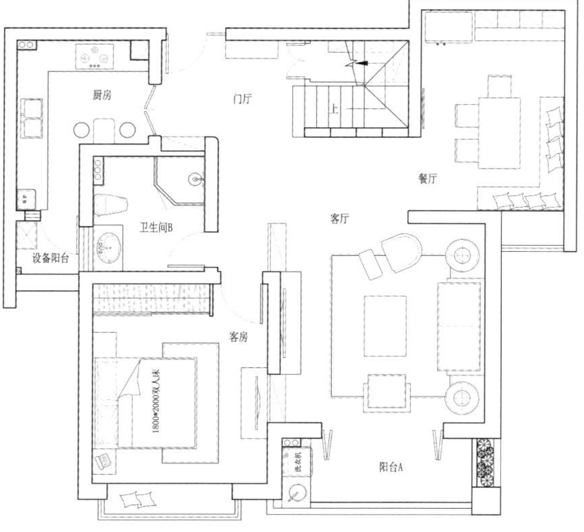

First Floor Plan / 一层平面图

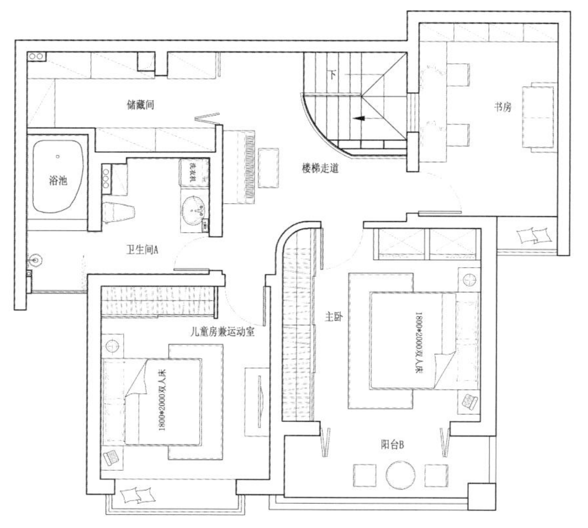

Second Floor Plan / 二层平面图

Cadres

COAST

New Canaan Residence

新迦南别墅

◎ Design Company: Specht Harpman Architects
◎ Designer: Specht Harpman
◎ Photographer: Elizabeth Felicella
◎ Location: Connecticut, America
◎ Area: 557 m^2

◎ 设计公司：Specht Harpman Architects
◎ 设计师：Specht Harpman
◎ 摄影师：Elizabeth Felicella
◎ 地点：美国康涅狄格州
◎ 面积：557 m^2

The New Canaan Residence is nestled into a lush hilltop site in this forested part of Connecticut. The house was designed to engage the landscape and immerse the occupants in the full range of environments that the site offers. The New Canaan Residence welcomes its owners and visitors into what feels like a floating pavilion in the tree canopy. A winding drive brings visitors through the forest to arrive at a low and gracious landscaped court that frames the glass entry pavilion. From this court, the transparency of the house is evident: floor-to-ceiling glass allows views through the house, incorporating the landscape into the body of the house.

The upper level features the home's primary gathering spaces, as well as bedrooms and a gymnasium. The lower level provides additional social spaces and features a media room, library and two home offices. The lower level spaces link to a series of "outdoor rooms" that feature a fireplace and distinct seating areas.

新迦南别墅坐落在康涅狄格州一座葱郁的小山顶部，别墅已深深地植入周围的环境之中。从远处看去，别墅似悬挂在树上的帐篷，正欢迎着主人和客人的到来。沿着蜿蜒的车道，穿过丛林，便可到达地势较低的观景庭院，由玻璃架构而成的观景庭院视觉通透，大幅的落地玻璃可欣赏户外美景的同时，也将室外的景色引入室内。

别墅的上层空间为家人聚会的场所，除此之外，还设有卧室和健身房。下层空间为社交空间，内有多媒体室、藏书室和两个家庭工作室。下层空间与户外相连，设有壁炉和会客区。

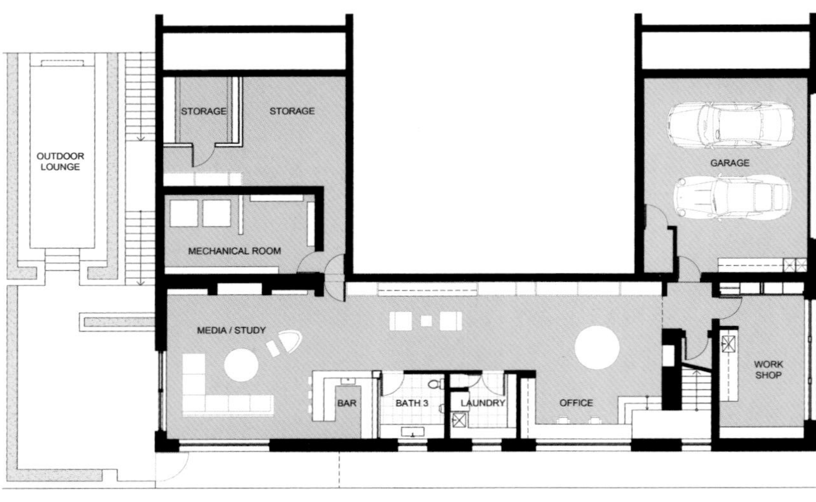

First Floor Plan / 一层平面图

Second Floor Plan / 二层平面图

West Lake Hills Residence

西湖山别墅

◎ Designer: Specht Harpman, Scott Specht, Louise Harpman, Sheryl Jordan, Brett Wolfe
◎ Photographer: Casey Dunn, Taggart Sorensen, Meg Mulloy, Scott Specht
◎ Location: Texas, America
◎ Area: 511 m^2

◎ 设计师：Specht Harpman、Scott Specht、Louise Harpman、Sheryl Jordan、Brett Wolfe
◎ 摄影师：Casey Dunn、Taggart Sorensen、Meg Mulloy、Scott Specht
◎ 地点：美国德克萨斯州
◎ 面积：511 m^2

On a densely tree covered site in the Austin, Texas exurb of West Lake Hills, Specht Harpman was tasked with the renovation and expansion of a modest 1970's house. Much of the original internal structure was maintained, but the alterations sought to erase all visible traces of the original house.

The primary goal of the new renovation and expansion was to preserve all of the site's large twisting Live Oak trees while creating a new and open home that embraces the landscape and weaves itself around the existing trees.

The primary mass of the house was lifted off the ground plane by creating a series of masonry walls that extend across the site to form interior and exterior rooms that frame views with the Live Oak trees beyond. The walls define family spaces on the lower level, while supporting a "floating box" of bedroom and private spaces above.

Ceilings of the existing house were low. For the expansion spaces, the ceiling plane was maintained, but the floor was lowered to follow the contours of the site, creating a terraced interior landscape that gives height to rooms while maintaining a modest exterior expression.

西湖山别墅坐落在美国德克萨斯州远郊的奥斯汀，被周围茂密的植被掩映，设计师 Specht Harpman 对这座 20 世纪 70 年代的别墅进行全面的改造和扩建后，室内结构并没有很大的改变，但设计也试图清除原建筑所有可见的痕迹。

在西湖山别墅改造和扩建过程中，设计师最初的目标是在保护周边环境和植被的基础上，打造一个全新的、开放的家庭空间，于此可以亲近自然，融入自然。

别墅内多数房间通过建造的砖石墙被提升至上层空间，大面的墙体限制了地势较低的家庭活动空间，同时，也支撑了上方做为卧室和私人空间的“漂浮箱”。

空间原有的天花很低，为了扩大空间，设计师并未拆除原有的天花，而是将地板移到了地平线的下方。同时，还新增了室内观景阳台，以此来增加室内的高度，维持别墅低调的外观。

GILBERT & GEORGE
GILBERT & GEORGE

OBEY

Norwegian Official Residence

挪威官邸

◎ Design Company: Dis. interiørarkitekter mnil AS
◎ Photographer: Sebastian Posingis
◎ Location: Colombo, Sri Lanka

◎ 设计公司：Dis. interiørarkitekter mnil AS
◎ 摄影师：Sebastian Posingis
◎ 地点：斯里兰卡科伦坡

Interior planning of representative areas in the Norwegian Official Residence, including hall, living room, meeting room, library, dressing room and dining room.

In addition to covering various functions of public visits the interior should present and emphasize a Nordic style. The designer had never worked with an official residence, and the client wanted to use this as an advantage and not put so many constraints so the designer would be able to see this with new eyes. The main objective of the Nowegian Official Residence in all of its projects is that most of the furniture, lighting and other essential elements shall promote Norwegian design.

The concept was designed around preserving the building's character and let the furniture and lighting be a contrast to emphasize the two different worlds that come together here. The colors on the furniture were chosen to emphasize the Scandinavian design and at the same time they were inspired by Sri Lankan traditional clothing and their rich natural landscape. Large rooms with dark wood and hard surfaces became the contrast to colorful furniture with a light and soft look.

挪威官邸的内部规划极具代表性，其内包括大厅、客厅、会议室、书房、更衣室和餐厅。

挪威官邸内部的装饰和铺陈极具北欧风格，因设计师从未设计过官邸，业主希望其不受任何限制，以新的视角来完成此次设计。本案设计的主要目标是通过家具、灯饰和其他基本元素来展现挪威设计。

本案的设计理念充分保留了建筑原有的风格，通过家具和灯饰的对比使两个完全不同的世界于此融合。家具突显了斯堪的纳维亚风格，与此同时，还受斯里兰卡传统服饰及地理环境的启发。大体量建筑坚硬的木质外表与室内浅色系的家具形成鲜明的对比。

Kowloon Bay Villa

龙湾别墅

◎ Design Company: Shangceng Voglass Decoration (Beijing) Co., Ltd. China
◎ Designer: Da Ming
◎ Area: 500 m²

◎ 设计公司：尚层装饰（北京）有限公司
◎ 设计师：大铭
◎ 面积：500 m²

The project is in a clear natural style, showing the love towards family and life. The integration of peaceful green, bright yellow, plain color is fabulous, creating a spring time in the house.

The light warm color, gentle and peaceful, reflecting the yearning for the nature. Actually, a designer need to not only pay attention to personal inspiration creation but also make the projects close to the life and the nature. Therefore, the designer merges the unruly reality into elegance to naturally form a style without ostentation.

As to the detailing, the designer avoids crudity and uses the perfect details to express personal housing taste. In addition to the visual effect and the aesthetic feeling, the utility and the functionality are also the viewpoint. The main bedroom is designed into an open space, emitting a breath of modern fashion. The shapes in the living room and the materials in the furnitures casually represent a feeling of comfort. The clear colors give rhythm to the vision and the American countryside style furnitures are glittering, embellished with the fabrics and flowers.

本案风格清新、自然，传达出女主人对家的理解和对生活的热爱。宁静、悦目的绿色，柔和、明快的黄色，还有淳厚、质朴的原色，鲜明大方，亦动亦静。住在这样的房子里，就如同生活在温暖的春天，永远地留住了欢愉。

本案在选色上偏向于轻松的暖色，柔和而宁静，体现了都市人对自然的向

KELLY HOPPEN HOME

KELLY HOPPEN HOME

往与追求。其实，作为一名设计师，在强调个性的同时，也需要使作品最大限度地贴近生活、回归自然。所以，在对这套别墅进行改造和装修的过程中，设计师并没有做过多张扬的奢华设计，而是将率性、真实融入其中，浑然天成，自成一派。

在细节的处理上，设计师极力避开简单、生硬的设计手法，从小处着手、大处落眼，让完美的细节表达独特的居家品位。在本案中，设计师不仅追求设计上的视觉效果与美感，对空间的实用性与功能性也细心考究。主卧被精心打造成一个开阔的空间，散发出现代、时尚的气息。客厅造型上的方与圆，材质上的棉与麻，在设计师的巧妙搭配下，呈现出随性的舒适感。在色彩上，大量清新、跳跃的色彩赋予视觉以律动感，偏向美式乡村风格的家具也在布艺和花朵的点缀下，熠熠生辉。

Casa LC

LC 别墅

◎ Design Company: Art Arquitectos
◎ Designer: Antonio Rueda
◎ Photographer: Sófocles Hernández
◎ Lighting Design: Noriegga Iluminadores Arquitectónicos
◎ Location: Mexico
◎ Area: 9,000 m^2

◎ 设计公司：Art Arquitectos
◎ 设计师：Antonio Rueda
◎ 摄影师：Sófocles Hernández
◎ 照明设计：Noriegga Iluminadores Arquitectónicos
◎ 地点：墨西哥
◎ 面积：9 000 m^2

A fantastic 9,000 square meters lot is the scene for the setting of the 2,000 square meters Casa LC. The house is located at Mexico City north in area where there are no high constructions and this advantage was used in the architectonic and landscape design incorporating to the project all the surrounding green areas.

The concept was resolved in 3 different areas. The social area was distributed around the grand central circular patio, which has a water mirror with irregular islets and a central stream. The perimeter circulation allows access to the different social spaces such as the vestibule with toilet, the library, the living room and the dining room, all linked to a hallway that goes around the patio.

The private area is formed by a square patio with an orange tree in the middle that is floating on top of islets inside of a water mirror. This magnificent space is surrounded by the 4 regular bedrooms and the master bedroom. All the bedrooms have an attic that goes out to the terraces in the second level that look to the square patio.

The entertainment area ends with the pool and has a swimming track, gaming room, home theater, gym, bathroom and a small house for guests. The bedroom attics are connected to the second level of the home theater

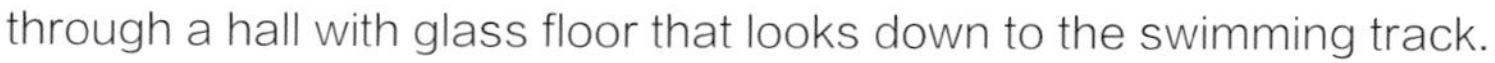

through a hall with glass floor that looks down to the swimming track.

These three sections are aligned with the main composition axis that joins the hallways and includes a semi private area formed by the TV room and the breakfast room. Here is also located the services are formed by kitchen, laundry and ironing room and the service rooms.

There are 4 sculptures located in the main ends of the house that represent the four natural elements: Earth, Fire, Water and Wind done by artist Carlos Marín. The lighting design has different variable scenes in a warm color, mostly done with LEDs, in a combination of indirect, accent and suspended lamps.

在墨西哥北部一处 9 000 平方米的地块上坐落着一栋 2 000 平方米的别墅——LC 别墅，周围并无很高的建筑。在建筑和景观的设计上，本案充分利用了这一优势，同时将周围的绿色植物也融入进来。

本案共分为 3 个不同的区域。社交区环绕在中央庭院的周围，庭院内的瀑布与不规则的岛屿和位居中央的小溪相得益彰。社交区的周围还配有卫生间、书房、起居室和餐厅，所有的功能区均环绕着庭院分布。

私人区由一个方正的庭院组成，内部的橘树漂浮在岛屿的中央。华丽的私人区被 4 个规则的卧室和 1 个主卧环绕，所有的卧室都有阁楼，登上二楼的阳台即可俯瞰整个中央庭院。

娱乐区位于水池的尽头，经由一条通往泳池的小径，即可到达。其内设有游戏室、家庭影音室、健身房、浴室和一间小小的客房。卧室的阁楼与二层的家庭影音室相通，通过大厅的玻璃地板即可看到下方通往泳池的小径。

本案的 3 个区域依空间主轴依次排开，主轴与门厅相连，其内包含一个半私密的空间，内设电视房和早餐台。与此同时，还设有厨房、洗衣房、熨衣房等服务区。

位于主建筑尽头的 4 个雕塑由艺术家卡洛斯 · 马林设计，代表了自然界的 4 种元素——土、火、水和风。以各种风景为主题的暖色调灯饰由 LED 制作而成，与吊灯遥相呼应，美不胜收。

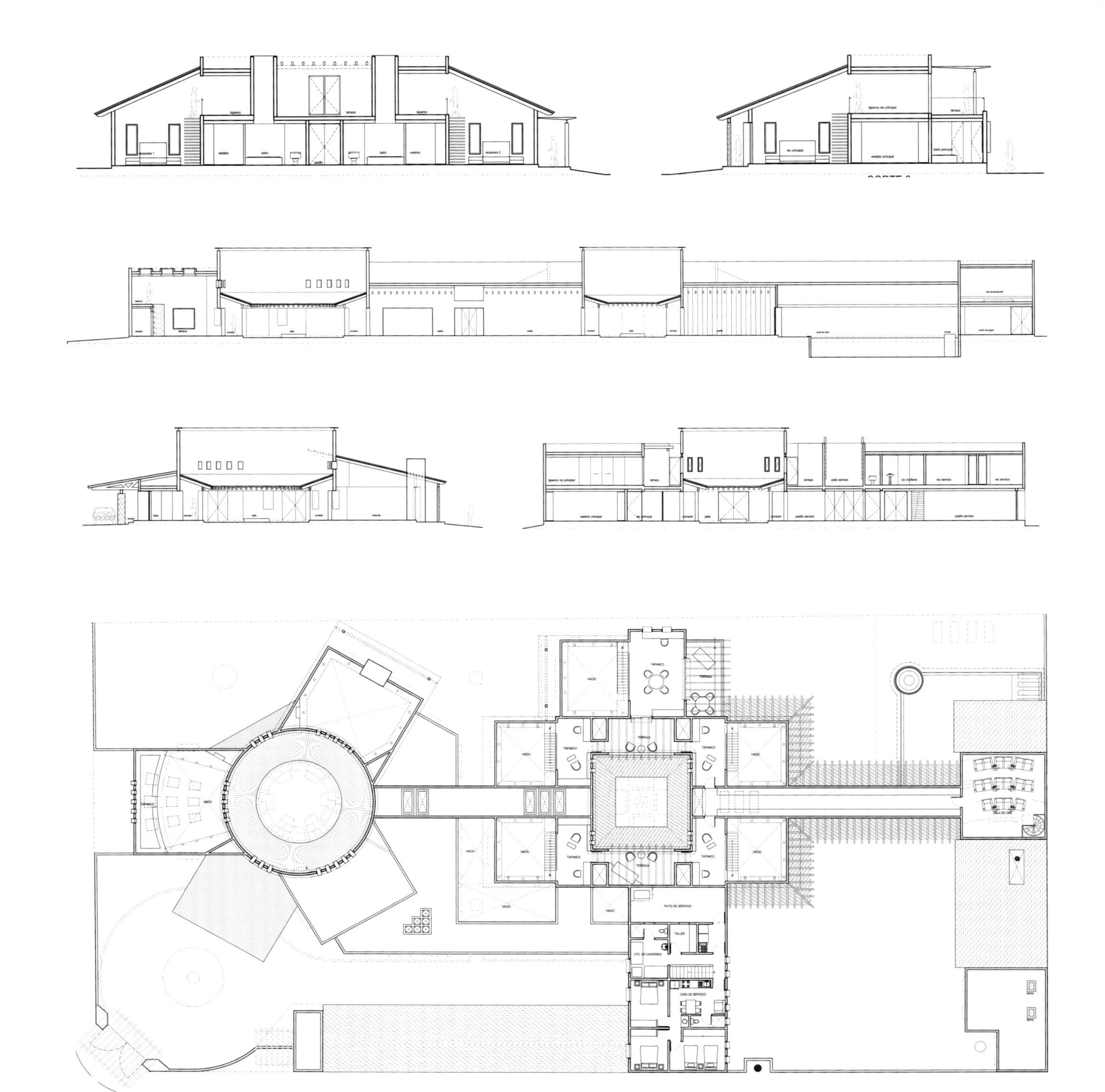

Site Plan / 平面图

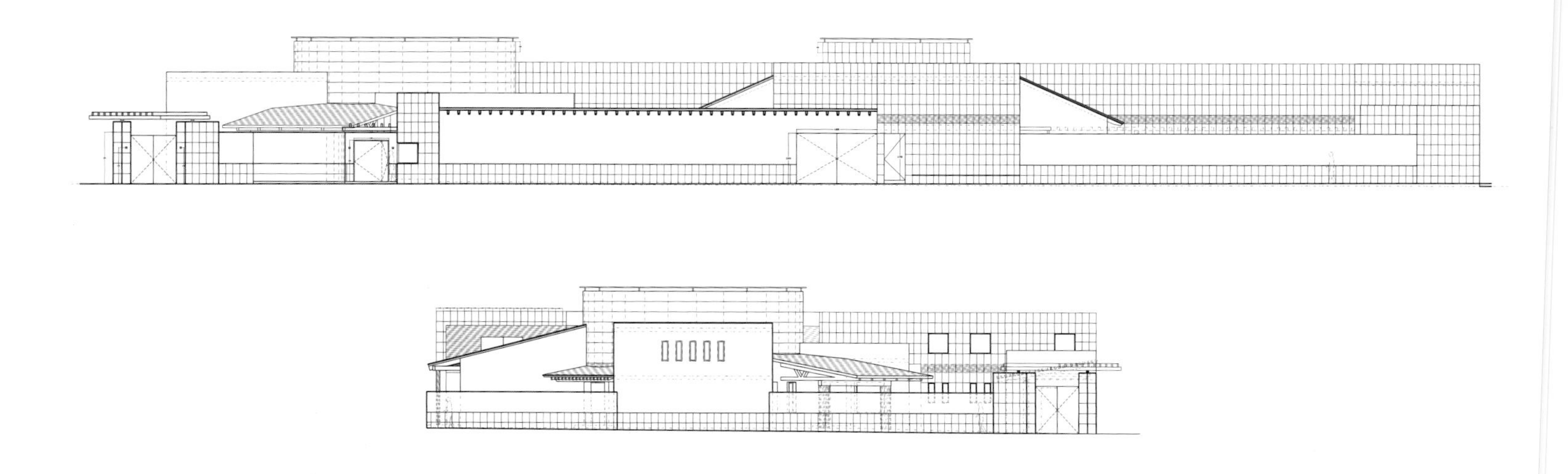

House in Rocafort

罗克福德别墅

◎ Design Company: Ramon Esteve Studio
◎ Designer: Ramon Esteve, Mercedes Coves, Sonia Rayos
◎ Photographer: Eduard Peris, Léa Chave, Silvia M. Martínez
◎ Location: Spain

◎ 设计公司：Ramon Esteve Studio
◎ 设计师：Ramon Esteve、Mercedes Coves、Sonia Rayos
◎ 摄影师：Eduard Peris、Léa Chave、Silvia M. Martínez
◎ 地点：西班牙

The beginning condition for this house is born from the trapezoidal shape of the plot, located in a residential area. The vegetation is a dense element of edge. Massive walls finished in bright white contrast the lush vegetation. Hidden by the trees, a wooden terrace takes the focus off the simple lines of the street facade.

Every room is provide with a different character. The inside-outside transition is really important to get it. The designer try to expand the space by extending the skin of the house to the outside. Besides, the empty spaces receive the light that shines through the holes. That works quite different depending on the time and the privacy degree. That is to say, the light shines bright and strong through the holes of the porch, veiled and controlled in the private spaces, getting to create dense atmospheres.

本案坐落在一个梯形的地块上，周围植被繁茂，大面积的白色墙体与葱郁的树木形成对比，隐藏在树木后面的木质阳台将视线从有简单线条的街道立面拉回此处。

每个房间的风格各异，室内外过渡自然。社交区与甲板和泳池相连，穿过玻璃门即可到达。外墙上镶嵌有大面积的木条板，为室内引入了更多的自然光。明亮的太阳光从条板间的空隙直接射入走廊，营造出浓厚的空间氛围。

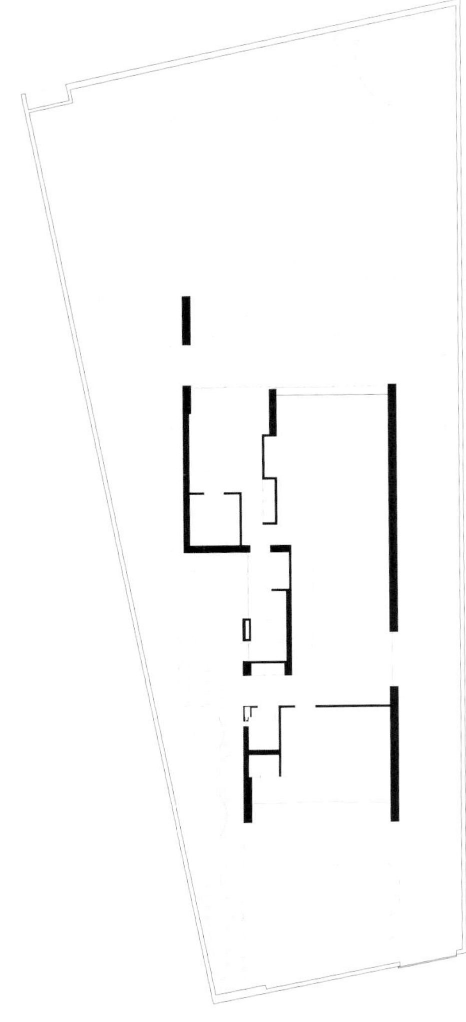

Xiamen Bali Villa

厦门巴厘香墅

- Design Company: East Design Decoration Eng. Co., Ltd.
- Designer: Zeng Guanwei
- Area: 500 m^2
- Main Materials: White Jade, Black Slate, Black and White Mosaic, Compound Floor

- 设计公司：东方设计装修工程有限公司
- 设计师：曾冠伟
- 面积：500 m^2
- 主要材料：白玉石、黑板岩、黑白马赛克、复合地板

Space

The project is a Bali style villa containing four levels including the underground space with a slope roof. The owner is a couple of elites of the generation after 80s who like the fashion and avantgarde modern style. The koi pond seperates the movie room and teahouse, the patio in the first floor becomes part of the interior seperating the living and dining area while integrating to meet the functional requirement, open and large. The patio in the second floor becomes a small half-open interior garden adjacent to the study, adding a touch of the nature. The space is clearly divided without too much shapes, clear and pure.

Color

The concise, warm and pure modern space expression is completed by the soft white jade, sedate black slate, fashional white and black mosaic and a warm wood color going through all the space, creating a timeless space. The most fantastic point in the space is that the gradual color changing red sofas in the living room echo with the greenery outside the window.

空间

这是一幢巴厘风格的现代别墅，含地下层共 4 层，业主是 80 后的精英，喜欢时尚、摩登的现代风格。地下层的观赏鱼池，隔开了休闲影院与品茶室，动静皆宜；一层天井的位置被设计成一个吧台，既分隔出客、餐厅，又在视觉上融为一体，整体空间一气呵成；二层的天井位置则变成一个半开放的室内小花园，紧邻书房，平添了些许自然的气息。空间区域划分清晰，没有多余的造型，干净纯粹。

色彩

设计师用温润的白玉石，沉稳大气的黑板岩，时尚摩登的黑白马赛克及延续所有空间的温暖木色，完成了对现代、简洁、温馨、纯粹的空间表情的塑造，四种简单的材料演绎出永不过时的黑白经典。客厅里渐变的红色沙发与窗外郁郁葱葱的绿意遥相呼应，是空间里最精彩的点缀。

Basement Plan / 地下层平面图

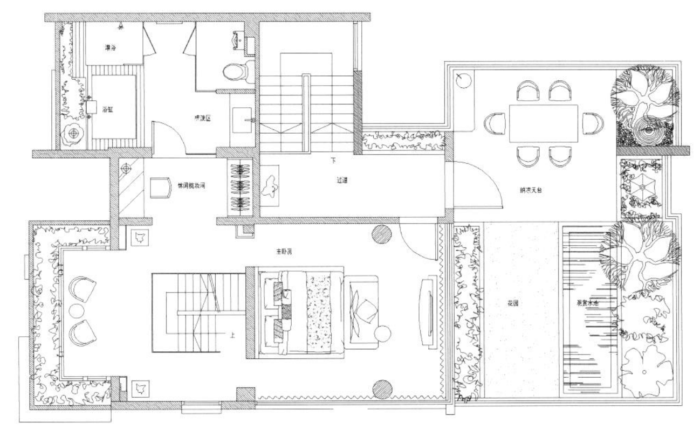

Third Floor Plan / 三层平面图

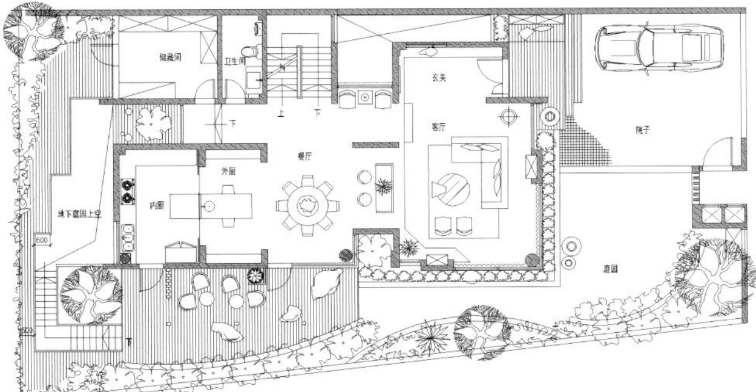

First Floor Plan / 一层平面图

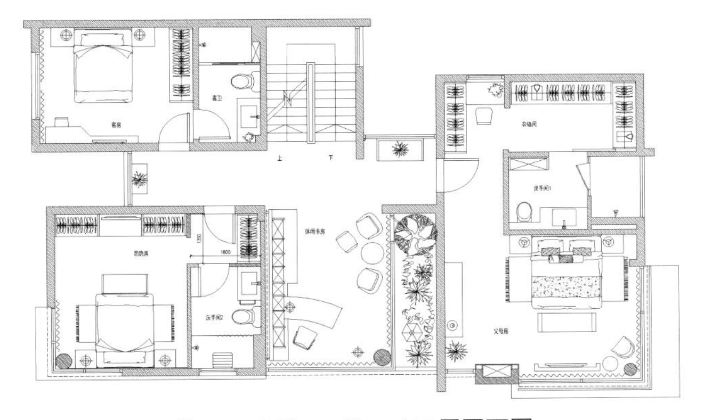

Second Floor Plan / 二层平面图

American Style FULL-SUN Villa

府尚别墅 · 美式

◎ Furnishing Design: Patrick Fong Design Ltd.
◎ Designer: Patrick Fong, Cowcow Cheng
◎ Area: 418 m^2

◎ 陈设艺术设计：方振华创意设计（杭州）有限公司
◎ 设计师：方振华、郑蒙丽
◎ 面积：418 m^2

The project is prepared for a big family. It is shined by the sunlight under which the communication make people more intimate.

Once you enter the house, you can see an elegant and magnificent house by the corner and gallery. The living room is huge and luxury, making the dining formal every time. The private area starts from the second floor, including a family room where the family can play and relax. The main bedroom in the third floor has a big terrace where people can appreciate the beautiful landscape outdoors. The children's room is also included in this level, benifitting the parenthood.

In the underground level, life becomes a point in the history, settled down in the visual items to creat the history of the family. There are wine cellar, wine cabinet, cigar bar, showroom, multi-media room, which all will become part of the wonderful life. For children, their childhood story will be stored here, carefree to grow up. One day, the child will enter the big world and get back to this reminiscent space like his father.

MARTELL
NOBLIGE

London Herald
London Herald

本案无疑是为一个大家族而准备的。多数房间采光良好，令整个空间充满明媚阳光，阳光下的沟通与交流升华着彼此的情感。

从进门开始，便能感受到这是一个大家族的居所，入户玄关与门厅的转角，就已经展示出大宅的典雅与气度。一楼挑空的客厅，气派十足，为主人营造出一种“高贵”的气氛，令每一次会客和用餐都拥有一种“仪式感”。真正的家庭私密空间是从二楼开始的，二楼还特别设计了一个家庭室，让家人在此体验温情、欢愉的美好时光。三楼主卧设计有环景大露台，男女主人每时每刻都能陶醉在室外的美景之中。同时，儿童房也设置在同一层，便于亲子关系的维护。

而在地下室，生活仿佛成了历史的一个个节点，凝固在了那些可见的物体上，用以开创这个家族的历史。在这里，男主人可以设置酒窖、酒柜、雪茄吧、陈列室、视听室……一杯醇酒的回味，一部经典影片场景的回眸，一件来之不易的古玩的回观，都将定格为美好生活的点点滴滴。对于孩子来说，这里将存放他们童年的故事——无忧的快乐，渐渐的成长。有一天，他也会走向更大的世界，也会像他的父亲一样，再度回到这里，沉醉于这个完全自我的空间。

Neoclassical Style FULL-SUN Villa

府尚别墅·新古典

◎ Design Company: Patrick Fong Design Ltd.
◎ Designer: Patrick Fong, Cowcow Cheng
◎ Area: 272 m^2

◎ 设计公司：方振华创意设计（杭州）有限公司
◎ 设计师：方振华、郑蒙丽
◎ 面积：272 m^2

This single family villa has four levels including the underground level. The first floor is staggered floor, lighting well, compact size and function well, with several entrance design, representing understated luxury and restraining.

Once entering, the luxury spiral stairs come into your eyes, along with the translucent crystal chandelier hanging down, magfinicent and elegant.

Entering into the living room from the southern side of the entrance, the 3.9 meters ceiling enhances the space, with the balcony bringing the landscape into the interior.

There is a about 100 square meters private entertainment space on the underground. In there we can see the movie room, the chess room, the bar, the boxroom, even the nanny room. People can have a good time in this space.

The two guest bedrooms in the second floor provide an excellent living environment with private bathroom, wardrobe and landscape balcony. The master bedroom in the third floor includes a cozy living room, an open door, a walk-in wardrobe, showing a unique taste and style with a "wasty" space.

这是一套相对独立的别墅，包括地下室共 4 层，一层错层设计，三面采光，面积紧凑，实用性强，还特别设计了多种入户方式，尽享低调奢华与内敛。

从玄关处入户，首先映入眼帘的是中庭左右对称的豪华旋转式楼梯，一盏晶莹剔透的水晶吊灯从上空垂下来，尽显美宅风范。

入户后，经由几级台阶，进入南向的客厅，3.9 米高的客厅空间，大大提升了室内的空间感，客厅外的多个一步阳台，引景入室，别有风情。

沿楼梯往下走，便是一个面积达百余平方米的地下私人娱乐空间。或可设置影视厅、棋牌室、红酒吧、收藏室等等，保姆房也可巧妙地设置在这儿，主宾可以在此尽情地享受生活的美好与欢娱。

二楼设置了两间客卧，均带有独立卫生间、衣柜及观景阳台，为家里的其他成员或来访者提供了良好的居家环境。而主卧位于三楼，舒适的起居室，双开大门，走入试衣柜，看似“浪费”的空间与尺度，却突显出主人不一般的品位与风格。

Spanish Style FULL-SUN Villa

府尚别墅 · 西班牙风格

◎ Design Company: Patrick Fong Design Ltd.
◎ Designer: Patrick Fong, Cowcow Cheng
◎ Area: 245 m²

◎ 设计公司：方振华创意设计（杭州）有限公司
◎ 设计师：方振华、郑蒙丽
◎ 面积：245 m²

For many people, it is very important to own an exclusive space such as an entertainment place for a hostess who has lots of friends. The first floor includes courtyard, dining room, living room which can be matched together freely, casually showing the clear life concept of the owner. Flowers appear from the table to the garden. The owner's great originality nearly can be part of the spring.

The home field is not just occupied by daily necessities, sometimes, it may be the continuation of life which can be seen in the dining room of FULL-SUN. The warm color and the exquisite tableware prepared by the hostess give the guest appetite and dignity. Various artstic Chinese and western dishes start the cooking sharing meeting cheerfully.

In addition to the comfortable and ebullient interior space, the exterior courtyard is also a best social place. The jagged, and orderly stair from the living room to the courtyard, and the landscape everywhere light the life.

Walking over to the courtyard, the fragrance comes with the breeze to comfort people, people enjoy the beauty like a good wine.

对于大多数人来说，有一个专属空间很重要。对于有着许多朋友的女主人来说，有一个心仪的待客场所尤为重要。这套房子的一楼拥有北院、餐厅、客厅、南内院等空间，不同的功能区域都可以很自在地混搭在一起，这样的空间组合看似随意，却充满了主人对敞亮生活的向往。从餐桌、茶几到花园，到处盛开着鲜花，女主人的独具匠心几乎可以成为春天的一部分。

家并不只是柴米油盐，有时，它也可能是社会生活的延续。府尚的餐厅完

全可以担此重任。阳光灿烂的上午，女主人早早地把自家的餐厅布置一番，暖暖的格调让人充满食欲，整套精致的餐具则给来宾带来赴宴般的尊贵感。各种艺术品般的中西菜肴，被女主人逐一端出，太太们的中西厨艺分享会，在一片惊呼声中开始了。

除了舒适、热情洋溢的室内空间外，室外的庭院也是很好的社交场所。从客厅到庭院，参差而有序的台阶，处处可见的景致，让生活的美好如生命般真实、生动。

沿着精致的石阶来到庭院，五月玫瑰的芳香随着一阵清风袭来，让人心旷神怡，沐浴其中，美好的生活情境在脑海中，如芳醇的美酒，渐渐氤氲开来。

Modern Style FULL-SUN Villa

府尚别墅 · 现代风格

◎ Furnishing Design: Patrick Fong Design Ltd.
◎ Designer: Patrick Fong, Cowcow Cheng
◎ Area: 324 m^2

◎ 陈设艺术设计：方振华创意设计（杭州）有限公司
◎ 设计师：方振华、郑蒙丽
◎ 面积：324 m^2

As an architecture designer, do not cater to the architectute but match the architecture with people's personality. It is more that the artists lived in FULL-SUN make FULL-SUN than that the charming of FULL-SUN conquers those people. Therefore, the designer decide to pay more attention to art space to match with the artists' characteristic.

For example, in order to highlight the artists' sense of presence, the designer start to create the space from the garage in the third floor underground. The bright and magnificent garage is designed to act as the transition to the famlily life area. A handmade chair, some pieces of elegant oil painting and the various featured items, not only show the owner's taste but also let the visitor start the plesant artistic life experience journey here.

There is a unique private colletion house in the first and second floor underground. In here there are a piece of woodcarving from Egypt and the various jewellery and old items. The private space is wonderful and artistic enough.

FULL-SUN pays more attention to the underground space: a wonderland for mind, art and soul.

作为一名建筑设计师，在做设计的时候，永远是让建筑来匹配人的气质，而不是要人来迎合建筑。与其说是府尚的建筑魅力征服了这样一些人，不如说是生活在这里的“艺术家”，成就了府尚。于是，在此次设计中，设计师决定将更多的精力，投入到艺术空间的设计上，以此来匹配居住在这里的“艺术家”的气质。

为了突出“艺术家”的“在场感”，设计师从地下三层——车库空间开始营造。车库层特别设置了罕有的、明亮且大气的家庭厅，作为停车空间向家庭生活空间的过渡，一张手工打造的椅子，几幅雅致的油画，以及独具特色的各种物件，无不体现出主人的高雅，也让来访者于此开始他愉快的艺术生活体验之旅。

在地下二层及地下一层空间，设计师为主人定制了一个绝无仅有的私人收藏馆。在这儿，可以存放主人不远万里从埃及带回来的一整条木雕，也可以展示女主人精心收藏的无数老首饰与老物件……在这片私人领域，足够私密，足够玩味，足够自由，当然，也足够艺术。

府尚，一个思想的栖息之地，一个艺术的酝酿之地，一个心灵的隐秘之地。府尚，专为这些人，寄存精神、理想和情怀。

Yihu Beautiful Home

怡湖 · 美家

◎ Design Company: North Rock Design
◎ Photography Company: Jin Xiaowen Space Photography
◎ Area: 280+60 m^2

◎ 设计公司：北岩设计
◎ 摄影公司：金啸文空间摄影
◎ 面积：280+60 m^2

Considering the function and aesthetic sense of the housing, the designer skillfully and innovatively creates a magnificent and remote life experience to enjoy a cozy and leisurely life through the usage of the natural and classical American casual style design elements. Let our life slow down to be enjoyed.

No matter the concise and clear line of furnitures or the old decorations show the comfort of life and simple fine life style.

American casual style make the home natural, elegant and magnificent, emitting its new charm in the city.

设计师采用自然、经典的美式休闲风格来设计本案，考虑到居家的实用性和美观性，设计师在设计上求新求巧，用空间本身营造出一种大气、久远的生活体验，享受慢生活的从容与舒适。

不论是简洁、明晰的家具线条，还是带有岁月沧桑的配饰，都在向人们展示生活的舒适和精致朴素的生活情趣。

美式休闲风格让家清新自然而不失优雅与华贵，当优雅与华贵融于一体的时候，美式的休闲风格便会在都市中焕发出新的魅力。

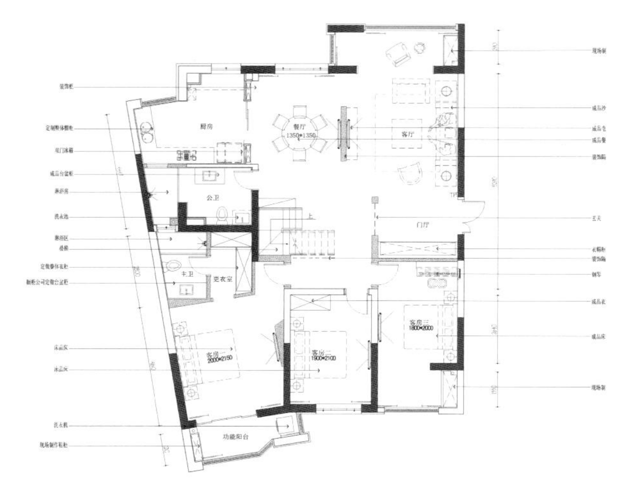

First Floor Plan / 一层平面图

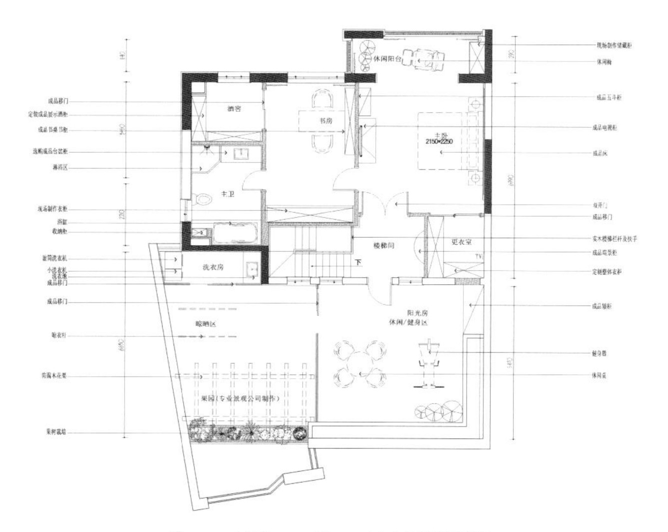

Seond Floor Plan / 二层平面图

Santa Monica Canyon Residence

圣塔莫尼卡峡谷屋

◎ Design Company: Griffin Enright Architects
◎ Designer: John Enright, Margaret Griffin

◎ 设计公司：Griffin Enright Architects
◎ 设计师：John Enright、Margaret Griffin

This residence is nestled into a hillside property in little Santa Monica Canyon. A long skylight extends the geometry of the path as it winds through the living space, illuminating the livings spaces with indirect light. The open, loft-like spaces of the residence are distinguished by the geometry of the meandering skylight. As the skylight bends, visually linking the front and back door of the house, it creates distinction among the kitchen, living area and dining area. The cross-section of the house is punctuated by the skylight, where it slips to further distinguish these living areas, shaping their individual volumes subtly within the whole. As it moves through the house, sculpting the high ceiling, the skylight connects the two new courtyards created by the form of the residence.

At the rear facade, pocket doors disappear to frame the view of the exterior from the living room. A deck at the rear views back on the living area and through the house to the front courtyard. The house is both held together as a whole, and divided into parts by the volumetric carving of the skylight and the exterior seems to pull through the house along the same path, as the visual continuity between the back and front courtyards is maintained along this zig-zag spine of the house.

此住宅坐落于圣塔莫尼卡峡谷山腰上。一道天窗循着几何路径蜿蜒穿过，间接照亮了整个居住空间，这是此住宅最显著的特点。天窗连接着房屋的前后门，将厨房、起居室、餐厅隔开。天窗强化了房屋的横断面，进一步区分出了起居空间，在统一的空间里巧妙塑造着每一个独特的区域。蔓延的天窗，雕饰着高高的天花板，连接了由房屋外形打造而成的两个新庭院。

屋后，打开起居室的一扇扇小门，室外美景映入眼帘。屋后的露天平台可以回望起居室，视线穿过房屋，可以看到前院。此住宅在维持整体统一的同时，也被天窗划分成不同的区域，外部有着相同的轨迹，如同房屋曲折的脊柱，使得前后庭院具有视觉上的连贯性。

Huizhou Zhongzhou Central Park

惠州中洲中央公园

◎ Design Company: KSL DESIGN(HK) LTD.
◎ Soft Installations: Shenzhen 1313 Decoration Co., Ltd.
◎ Designer: Aday Lam, Wen Xuwu
◎ Location: Huizhou, Guangdong
◎ Area: 980 m^2

◎ 设计公司：KSL 设计事务所
◎ 软装设计：深圳壹叁壹叁装饰有限公司
◎ 设计师：林冠成、温旭武
◎ 地点：广东惠州
◎ 面积：980 m^2

The walls made of handmade red brick, having a concise line and heavy colors, exude a natural breath. In the respective of interior decoration, the living room show a magnificent space with the elegant crystal chandelier. Every luxury furniture emits the charming of the interior space, perfect in the function and style.

本案墙体选用了手工打制的红砖，线条简洁，色彩凝重，流露出自然的气息。在室内装饰上，客厅用高雅的水晶灯彰显空间的大气，每一件奢华的家具都散发着空间内在的魅力，在功能与风格的把握上恰到好处。

三國演義
三國演義
三國演義
三國演義

三國演義

POST

Candy

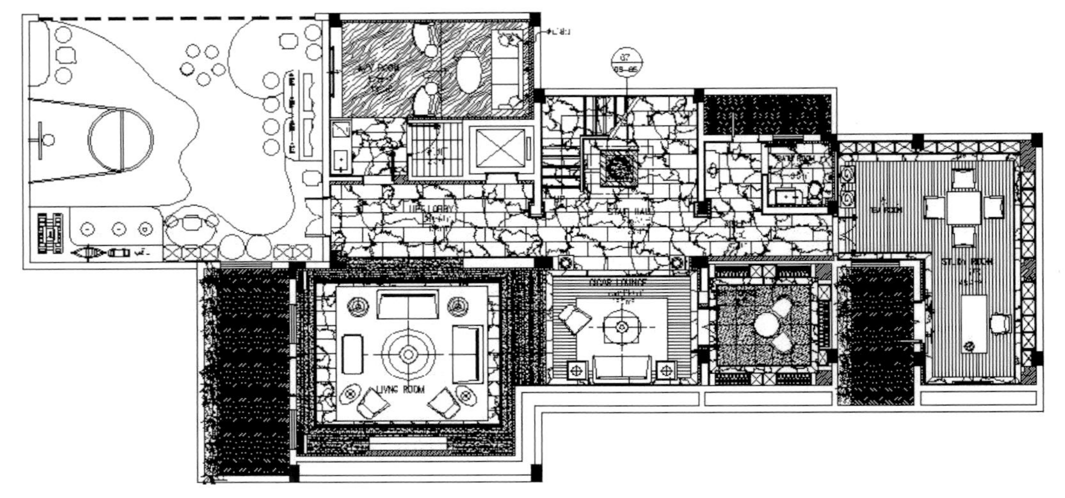

Basement Plan / 地下层平面图

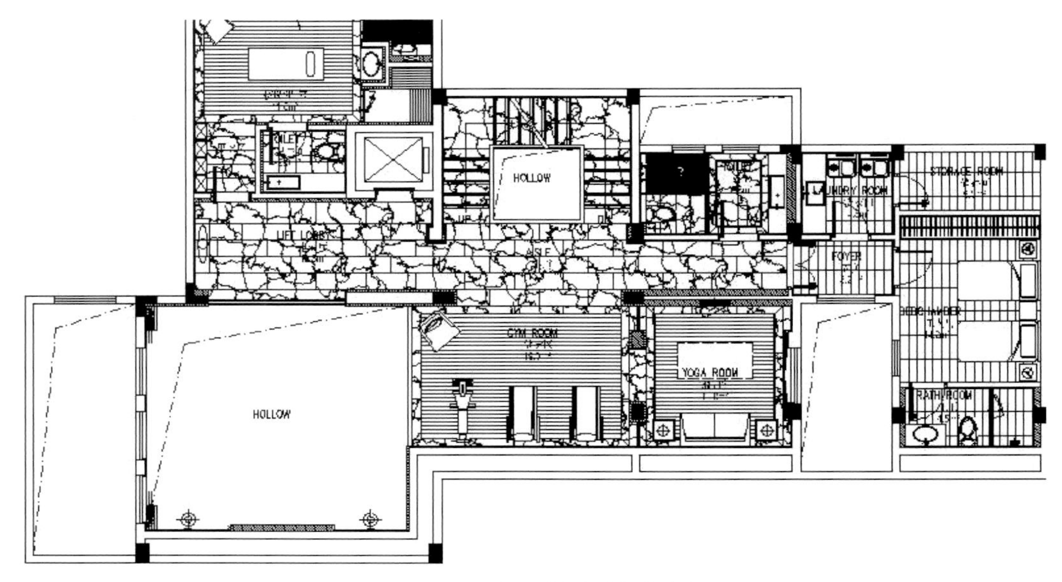

Basement Plan / 地下层平面图

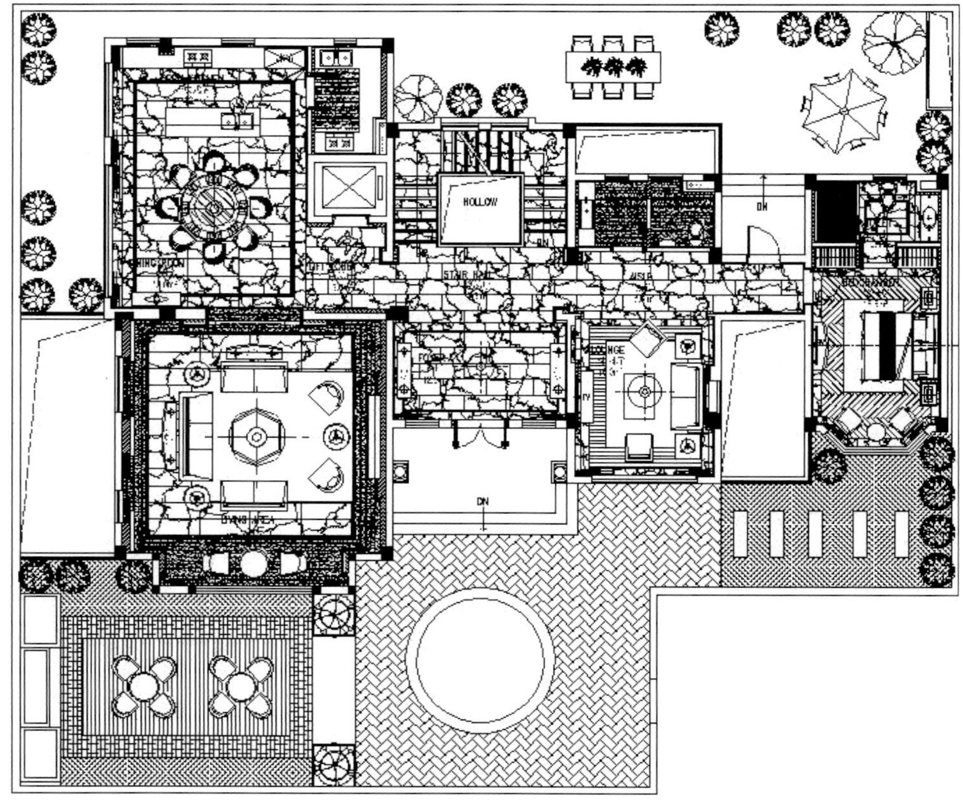

First Floor Plan / 一层平面图

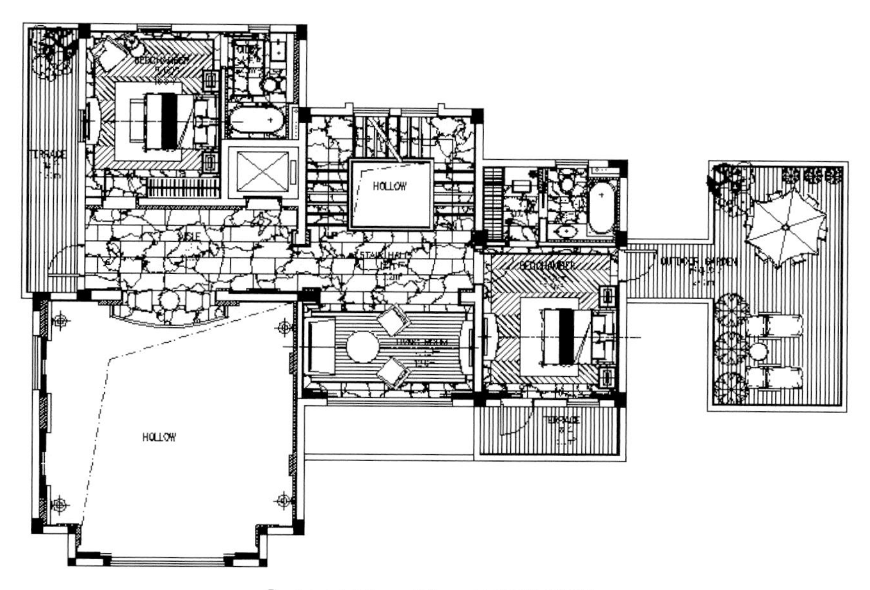

Second Floor Plan / 二层平面图

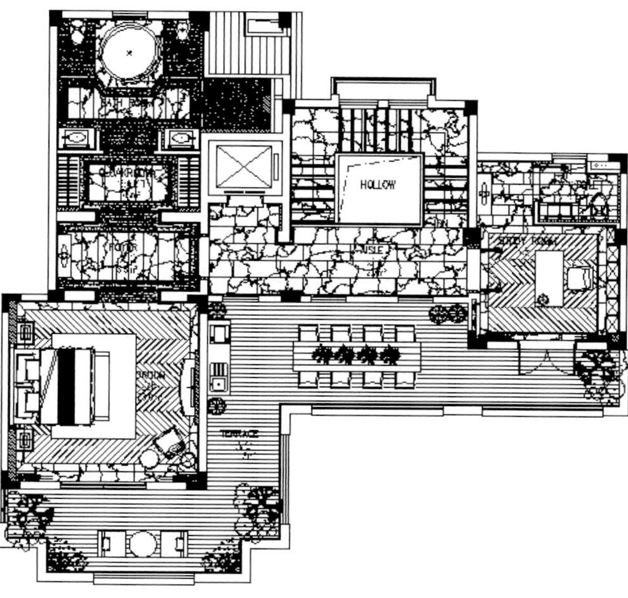

Third Floor Plan / 三层平面图

紅樓夢

West Hill Whispering Woods

西山林语

◎ Design Company: Shangceng Voglass Decoration (Beijing) Co., Ltd. China
◎ Designer: Zhang Ping
◎ Area: 450 m^2

◎ 设计公司：尚层装饰（北京）有限公司
◎ 设计师：张萍
◎ 面积：450 m^2

The project is a 450 square meters countryside villa. The owner collects plenty of classic furnitures, having a knowledge of the Chinese traditional culture, fond of scripts and paintings. However, the owner thinks that Chinese style furnitures lack comfort and colors in the housing.

The project is inspired by the communications between the "courtyard" in Chinese style traditional architecture and every space. The nearest light well is converted into a sunlight room which becomes a vivid highlight through the conatation of "water, tea, music and sunshine".

The space is easy, cozy, modern and concise, where the modern furnitures still play an important role, as well as the natural Chinese style classic furnitures accented with exquisite color application, decorations and paintings.

本案是一幢450平方米的乡间别墅，业主收藏有大量的古典家具，在中国传统文化上有一定的造诣，业余时间喜欢品鉴字画，但认为中式家具色彩单一，且缺乏舒适感。

本案的设计灵感来自中式传统建筑的“庭”与各个空间的和谐沟通。与庭最亲近的下沉采光井被改造成阳光房，水、茗、音乐、阳光成为空间亮点。

各功能空间则保持轻松舒适、现代简约的基调，空间中的现代家具仍是主角，整体空间中精妙的色彩、配饰、画作使点缀其中的中式古典家具显得自然而不突兀。

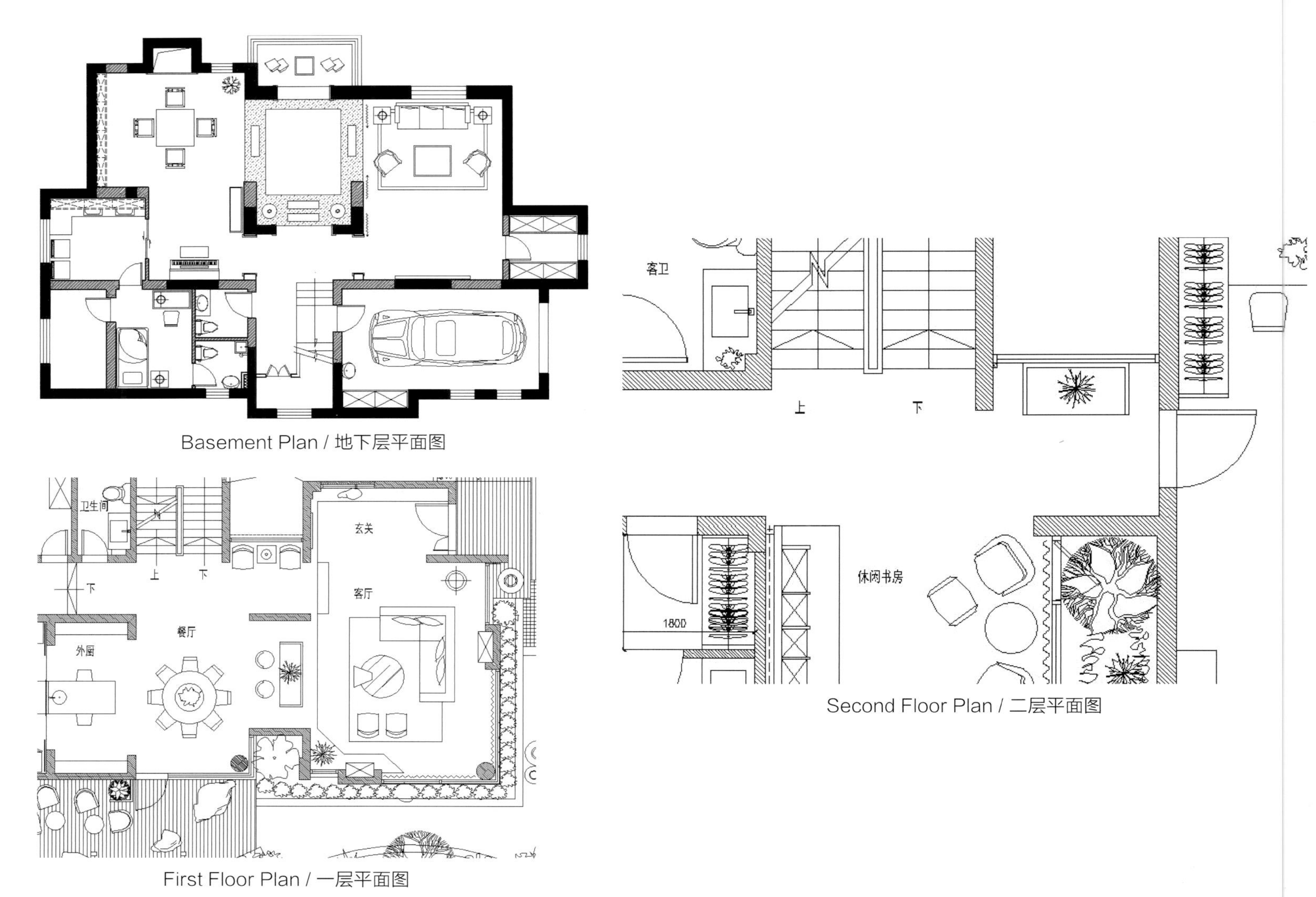

Basement Plan / 地下层平面图

Second Floor Plan / 二层平面图

First Floor Plan / 一层平面图

Chinese Tuscan Dream

中国人的托斯卡纳梦

◎ Designer: Laogui
◎ Location: Zhengzhou, Henan
◎ Area: 320 m^2

◎ 设计师：老鬼
◎ 地点：河南郑州
◎ 面积：320 m^2

The ivory chalk, the gold Tuscan sunshine, the red soil, the green forest, the vineyard, the rangeland, the simple furniture, emit casual and comfortable countryside breath, inlaid in the green field like a graceful peotry, enhanced with the dark red gem, chinati wine and tomatoes, creating the Tuscan.

Log suspended ceiling merges with the arch, is fantastic like a huge oil painting. Surroundded by the landscape outside, the housing provides a Tuscany peaceful and original ecological life style. Waken by the birds' singing, one can enjoy a plesant life walking over the trees and pools.

休闲、舒适的乡村气息，古朴的家具，象牙白的白垩石，金色的托斯卡纳阳光，红色的土壤，葱郁的森林、葡萄园和牧场，浅绿色的橄榄果园，犹如优美的田园诗一般镶嵌在这片绿色的山野上，更有深色的红宝石，光泽的基安蒂红葡萄酒和鲜红的番茄……各种颜色调和在一起就是托斯卡纳。

设计将拱门与原木吊顶结为一体，宛如一幅硕大的油画，耐人寻味。依托室外景观而建的私家宅院，在蓝天白云下缔造出托斯卡纳般安详、原生态的生活。树荫盈地，池水清澈，被清晨的鸟鸣唤醒，沿着青石板信步，细细品味生活中的乐趣……

A16 Villa Show Flat, Nansha Xinghe

星河南沙 A16 别墅样板房

- Design Company: Matrix Design
- Area: 600 m^2
- Main Materials: Jazz White, Peacock King, Solid Wood Floor, Grey Mirror, Black Mirror

- 设计公司：矩阵纵横设计
- 面积：600 m^2
- 主要材料：爵士白、孔雀金、实木地板、灰镜、黑镜

The project uses the concept of "four water to the court" to bring the garden's water system inside, with the method of stream and waterfall going through the two levels, enhancing the set-in luxury villa concept. At the same time, the view borrowing is charming every level. It creates a high-end space.

本案在入户时便开始用“四水归堂”的概念将整个小区的水系引入室内，并将流水、瀑布的设计手法广泛应用于两层空间，在空间关系上强调了“登堂入室”的豪宅理念。同时，层层借景，引人入胜。在一个原本只是洋房的底层空间打造出一个可与别墅媲美的高品质空间。

OLD WORLD INTERIORS

New World Dream Lakeside Villa

新世界 · 梦湖香郡

- Design Company: Wuhan Zhang Jizhong Interior Architect Studio
- Designer: Zhang Jizhong
- Area: 485 m^2
- Main Materials: Clapboard, Marble, Stone Mosaic, Antique Brick, Solid Wood Floor

- 设计公司：武汉张纪中室内建筑
- 设计师：张纪中
- 面积：485 m^2
- 主要材料：护墙板、大理石、石材马赛克、仿古砖、实木地板

Situated in the Tower Lakeside and Dream Lakeside in the center of Wuhan, positioning in a customized top residence, the project creates the omnipresent feeling of dignity and honor for the elites in the city.

In the respect of facades, decoration elements, fabrics and other ornament, the villa largely uses the longitudinal perpendicular lines to enhance the depth and dimension sense of the space. The bright interior space merges with the beautiful outdoor scenery in the lakeside.

The main colors are white, light green and light brown, the different levels in the same colors create spatial level and third-dimension by shadows. Some bright colors, plants and crystal ornaments enliven the atmosphere. The main elements runing through the whole space consist of white wood veneer, light bright floor and furnitures, light green curtains, oil paintings, fabrics and wallpaper. The concise and exquisite detailings relax and please people.

Because of the open and bright space, the first floor involves the hall, the living room, the kitchen and other public area. The private area is designed in the second floor, including the living area and children's room, adding a touch of softness and peace accented by the furnitures, fabrics and decorations. The master suite shows the understated exquisite personal taste by the American style furnitures in deep colors with the endless scenery and terrace outside.

本案坐落于武汉城市中心区的塔子湖与梦湖公园之畔，是为富庶之家量身定制的领袖级豪华别墅，让创造城市的才智名流感受无处不在的尊崇。

别墅在建筑立面、装饰元素以及布料等配饰的选择上大量使用竖线条，利用序列线条增加空间的纵深感，从而提升空间的体量。室内空间光线充足，敞亮通透，与优美的室外美景融为一体。

空间主色调控制在白、浅绿、浅棕之间，同一色系不同色阶之间通过阴影营造出空间立体感，光影更加丰富。局部少量鲜艳色彩、绿色植物以及水晶制品的点缀，活跃了空间的整体气氛。白色的木质饰面、浅棕色的地板以及桌椅、浅绿色的窗帘、油画、布艺以及壁纸，构成了贯穿整个别墅空间的主体元素，不饰浮华却又简洁、精致的细节，让人的审美感官松弛而又敏锐。

由于空间高大通透，光线充足，一楼设置了门厅、客厅、餐厅、厨房以及临湖露台等公共空间。二楼属于家庭的私人领地，在起居室、次卧、儿童房之中，设计师融入了几分温馨与柔和，在线条优雅的家私、软装布艺以及饰品的衬托下，更加柔和、安逸。主人房采用了套房的形式，设计师精心挑选了美式风格的家具，借以表达业主精致、低调的品位。主人房也以深色为主，窗外的宽敞露台以及窗帘后的无限风景，给人尊贵的感觉。

Vuitton Town of Hot Springs Villa-B

威登小镇温泉 Villa-B

◎ Design Company: Symmetry Design
◎ Decoration Design: MoGA Decoration Designt
◎ Designer: Calvin Chen, Amy Lee
◎ Location: Weihai, Shandong
◎ Area: 161 m^2

◎ 设计公司：大匀国际空间设计
◎ 软装设计：上海太舍馆贸易有限公司
◎ 设计师：陈亨寰、李巍
◎ 地点：山东威海
◎ 面积：161 m^2

There is a quiet and elegant private villa in a solid and plain lime wall. This villa includes terrace, swimming pool and winding streams...

The designer uses natural landscape to create private space featured by "without din of horse and carriage". The villa has excellently geographical environment featured by free from hurly and near mountain and water. No matter in living room, dining room or Japanese stream room, ingenious application of natural material such as bamboo and stone without polishing forms the whole style trend. The designer aims to create a kind of living environment which is natural and has full personality.

In the living room, the leisure texture built by original stone skin echoes the thin column ceiling with black-dyeing by walnut vertically. Besides, living room is matched with lively design furniture. In this way, the art space full of modern sense is completed.

The connection between living room and bar forms broad space sense. The independent restaurant space closely echoes the landscape. The owner even can have visual feeling of open and concise large private swimming pool and elegant outdoor landscape.

The Japanese stream room and tatami style teahouse outline relaxed and enjoyable atmosphere. Besides, they create leisure and classical sense which is like Japanese traditional hotel. The bath pool which is collaged by weed tree brings fresh and carefree extreme enjoyment; the warm tone timber veneer is matched with cold tone stone with gray wood grain so that it adds coordinative and comfortable space feelings.

Ingenious configuration of double master bedrooms and indoor Japanese stream room add negotiability and layering to the space. During the sunset, hold a cup of wine and lie down on the terrace. In the grand and quiet atmosphere, smile at each other in a most beautiful place, enjoy the beauty holding in the heart.

厚重朴实的石灰墙里是一幢静谧雅致的私人别墅，露台、泳池、曲水流觞……

设计师利用自然景观打造“而无车马喧”的私享空间。别墅拥有远离喧嚣、依山伴水的绝佳地理环境。无论客厅、餐厅、日式汤屋的设置，还是竹木和石材等天然材质的巧妙运用，都形成了空间的风格趋向，意在打造一处自然且富有个性的生活环境。

客厅以原石皮营造出休闲质感，与胡桃木染黑的细柱屋顶垂直呼应，衬以风格明快的家具，使空间现代感十足。

客厅与吧台的连接使空间更显宽敞，独立的餐厅空间与景观亲密呼应，于此可直观空旷简约的大型私人泳池与雅致的户外美景。

日式汤屋与榻榻米风格的茶室，轻松勾勒出写意的氛围，营造出仿如日本传统旅馆的悠闲、古典质感。使用杂木拼贴而成的泡汤浴池，给人清新、畅快的极致享受；暖色调的木饰面搭配冷色调的灰木纹石头，营造出协调、舒适的空间感受。

双主卧空间与室内日式汤屋的巧妙搭配，更增加空间的流通性及层次感。夕阳西下之时，捧一杯美酒，卧于露台间，在静谧的气氛里，于风景绝佳处，相视一笑，感受沁人心脾的美好。

Vuitton Town of Hot Springs Villa-C

威登小镇温泉 Villa-C

◎ Design Company: Symmetry Design
◎ Decoration Design: MoGA Decoration Designt
◎ Designer: Calvin Chen, Zakia Zhang
◎ Location: Weihai, Shandong
◎ Area: 169 m^2

◎ 设计公司：大匀国际空间设计
◎ 软装设计：上海太舍馆贸易有限公司
◎ 设计师：陈亨寰、张三巧
◎ 地点：山东威海
◎ 面积：169 m^2

SNAPS

As a resort villa, whose theme is female, the project aims to show the housing beauty which is as gentle and lovely as the female, therefore, the whole space is pure white, simple and elegant, adding a touch of peace and grace. A large area of landscape windows result in a transparent space, showing the taste of the owner in an open way, echoing with the outdoor courtyard and the lakeside scenery. Such a layout brings more sceneries into the house and naturally divides the fuctional areas.

The white curtains and elegant furnitures in the living room echo with the relaxed holiday atmosphere. The designer symmetrically takes use of the traditional concept and the modern style, along with the suspended ceiling, creating a nature space.

The breakfast table in the dining room can also act as the working area, resulting in an open space, where the family can have more fun in life.

A light-colored glass divides the bedroom and the bathroom, ensuring the openness and the privacy. The plain-colored fabrics, the elegant furnitures and the soft installations in the bathroom light the space in the lamplight, providing people with a qualified life.

The scene can only be depicted by a line of Xu Zhimo, "The intangible stream, is waken and surprised by the brilliant cloud which is dancing lightly and moving gracefully, thus to hold the shadow of the cloud tightly though it itself is humble."

本案是一套以女性为主题的度假别墅，意在展现如女性般柔美如水的居室之美，因此，空间的整体色调以纯白的素雅之色为主，让空间显得宁静而优美。在空间的布局上，设计师通过大面积景观窗的设置使空间通透，以一种开放的姿势展现主人的审美与阅历，更与室外春意盎然的前庭、后院及美丽的湖景相映成趣。这样的布局，视野开阔，目光所及之处无不纯净；衔接处的微处理又自然地将功能区进行划分开来。

客厅运用素白的窗帘和雅致的家私，来营造轻松、柔美的度假感。设计师将传统意境和现代风格融合，配以精心设计的吊顶，更享自然的审美意境。

餐厅的早餐台兼作工作区，使空间连贯一体，也更适合一家人甜蜜的享受生活的乐趣。

卧室和洗浴间通过浅色的玻璃隔开，整个空间既相通又独立，开放性和私密性达到完美结合。素色的布艺，高雅的家私，以及精心打造的洗浴软装在灯光的明暗之间一张一合，让空间错落有致，极具生活品质。

此情此景，或许也只有徐志摩的那句“有一流涧水，虽则你的明艳”能诠释吧！

Cang Hai Villa No. 1 Townhouse Unit A1

苍海一墅联排别墅 A1 户型

- Design Company: Chongqing Pin Chen Design
- Designer: Pang Fei, Zhang Yan
- Location: Dali, Yunnan
- Area: 254 m^2
- Main Materials: Rainforest Brown Stone, Solid Wood, Rust Yellow Stone

- 设计公司：重庆品辰设计
- 设计师：庞飞、张雁
- 地点：云南大理
- 面积：254 m^2
- 主要材料：雨林棕石材、杉毛榉实木板、锈石黄石材

CINDERELLA
CINDERELLA

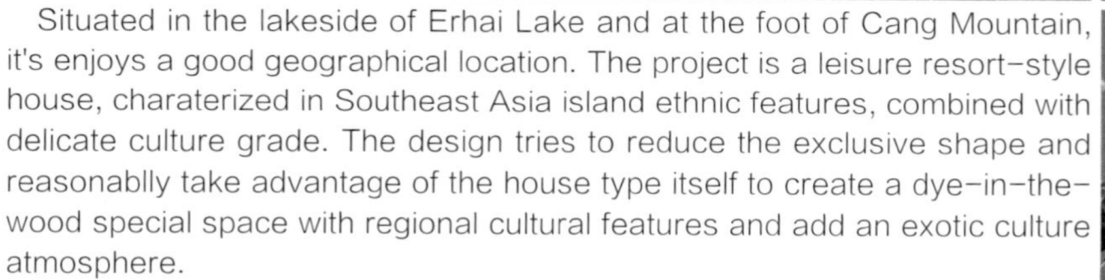

Situated in the lakeside of Erhai Lake and at the foot of Cang Mountain, it's enjoys a good geographical location. The project is a leisure resort-style house, charaterized in Southeast Asia island ethnic features, combined with delicate culture grade. The design tries to reduce the exclusive shape and reasonablly take advantage of the house type itself to create a dye-in-the-wood special space with regional cultural features and add an exotic culture atmosphere.

The materials widely involve the high-quality wood and natural marble, accented by rattens, bamboos and dark furnitures. The unique humanistic amorous feelings show itself in the project, which represents the perfect life experience given by the tourism real estate.

本案地处云南大理的洱海边、沧山脚，地理位置优越。本案别墅做为休闲度假之所，具有东南亚民族岛屿风情，又与其精致的文化品位相结合。在此次设计中，设计师力求减少过度的造型，合理地使用户型本身的优势，着重打造风味十足、充满地域特色的人文空间，营造出异域文化氛围。

在材质的选择上，本案广泛运用了大理本地的优质木材和天然石材，搭配藤条、竹子、深木色的家具等。本案的独特人文风情由此可见一斑，体现了旅游地产给人的完美生活感受。

RACING
FLOTT

The Feeling of "Play in the Art"

游于艺的 Feeling

◎ Designer: Laogui
◎ Location: Shenzhen, Guangdong
◎ Area: 533 m^2

◎ 设计师：老鬼
◎ 地点：广东深圳
◎ 面积：533 m^2

The feeling of "Play in the Art", the newest work of the designer Laogui, tells us that what is the highest realm of the real "Play in the Art" using the earthiest method to express the most sincere emotion.

In the project, the designer proposes to embellish every detail with fun and joy, imperceptibly showing the contral for the whole space. The paving in the living room stays the same as the color of the furniture, flexibly lighting the whole space.

The decorations keep in form with the style of the space, natural and not reserved. The design of the stairs takes advantage of the method of "break before create" to intersperse every step with national characteristic pattern tiles, spotted by the side lighting as well as the design ofthe banisters, completely corresponding to the theme of "Play in the Art". In the space transition, the designer continues the emotional design concept to design skillfully in color and

shape, especially the suspended ceiling which is fantastic and naturally designed. The spot in every room focuses on the background of the bed which becomes the visual centre of the viewer.

本案设计师老鬼的最新作品“游于艺的 Feeling”，以最朴实的手法，最真实的情感，告诉我们什么才是“游于艺”的最高境界。

在本案设计中，设计师主张以生活情趣来点缀空间的每一个细节，无形中透露出空间的主人翁意识。客厅的铺贴色调与家具的色系保持一致，灵活的铺贴方式，使整个空间不显单一。

在软装的设计上，设计师沿用空间原有的风格，自然而不拘谨。楼梯处，采用先“破”而立的设计手法，用民族性的图案瓷砖来点缀每一级踏步，同时辅以侧面的灯光，以及楼梯扶手的“破”，使楼梯成为空间的亮点之一，完全符合“游于艺的 Feeling”的设计主题。在空间过渡的处理上，设计师始终秉承情感设计的理念，无论是在色彩还是造型的处理上，设计师都如鱼得水。吊顶的处理更是精彩自然，不拘一格，没有丝毫的强求。每一间卧室的亮点都集中在床头的背景墙上，成为人们的视觉中心。

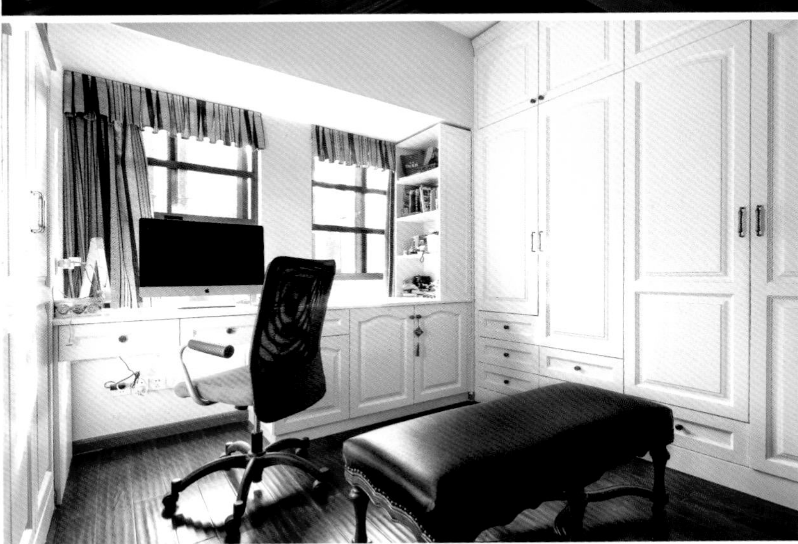

图书在版编目（CIP）数据

臻藏美墅 / DAM 工作室 主编 . – 武汉 : 华中科技大学出版社 , 2013.9

ISBN 978-7-5609-9390-4

Ⅰ . ①臻… Ⅱ . ① D… Ⅲ . ①别墅 – 建筑设计 – 世界 – 图集 Ⅳ . ① TU241.1-64

中国版本图书馆 CIP 数据核字（2013）第 233873 号

臻藏美墅 DAM 工作室 主编

出版发行：华中科技大学出版社（中国・武汉）
地　　址：武汉市武昌珞喻路1037号（邮编：430074）
出 版 人：阮海洪

责任编辑：王莎莎　　责任监印：张贵君
责任校对：熊纯　　装帧设计：筑美空间

印　　刷：中华商务联合印刷（广东）有限公司
开　　本：965 mm x 1270 mm　1/16
印　　张：21.25
字　　数：170千字
版　　次：2014年1月第1版 第1次印刷
定　　价：350.00元（USD 69.99）

投稿热线：（020）36218949　　1275336759@qq.com
本书若有印装质量问题，请向出版社营销中心调换
全国免费服务热线：400-6679-118 竭诚为您服务